Anna Taubenberger

Quantifying cell-ECM adhesion by single cell force spectroscopy

AF092596

Anna Taubenberger

Quantifying cell-ECM adhesion by single cell force spectroscopy

Quantifying adhesive interactions between cells and extracellular matrix by atomic force microscopy based single cell force spectroscopy

Südwestdeutscher Verlag für Hochschulschriften

Impressum/Imprint (nur für Deutschland/ only for Germany)
Bibliografische Information der Deutschen Nationalbibliothek: Die Deutsche Nationalbibliothek
verzeichnet diese Publikation in der Deutschen Nationalbibliografie; detaillierte bibliografische
Daten sind im Internet über http://dnb.d-nb.de abrufbar.
 Alle in diesem Buch genannten Marken und Produktnamen unterliegen warenzeichen-, marken-
oder patentrechtlichem Schutz bzw. sind Warenzeichen oder eingetragene Warenzeichen der
jeweiligen Inhaber. Die Wiedergabe von Marken, Produktnamen, Gebrauchsnamen,
Handelsnamen, Warenbezeichnungen u.s.w. in diesem Werk berechtigt auch ohne besondere
Kennzeichnung nicht zu der Annahme, dass solche Namen im Sinne der Warenzeichen- und
Markenschutzgesetzgebung als frei zu betrachten wären und daher von jedermann benutzt
werden dürften.

Verlag: Südwestdeutscher Verlag für Hochschulschriften GmbH & Co. KG
Dudweiler Landstr. 99, 66123 Saarbrücken, Deutschland
Telefon +49 681 37 20 271-1, Telefax +49 681 37 20 271-0
Email: info@svh-verlag.de
Zugl.: Dresden, TU, Diss., 2009

Herstellung in Deutschland:
Schaltungsdienst Lange o.H.G., Berlin
Books on Demand GmbH, Norderstedt
Reha GmbH, Saarbrücken
Amazon Distribution GmbH, Leipzig
ISBN: 978-3-8381-2427-8

Imprint (only for USA, GB)
Bibliographic information published by the Deutsche Nationalbibliothek: The Deutsche
Nationalbibliothek lists this publication in the Deutsche Nationalbibliografie; detailed
bibliographic data are available in the Internet at http://dnb.d-nb.de.
 Any brand names and product names mentioned in this book are subject to trademark, brand
or patent protection and are trademarks or registered trademarks of their respective holders.
The use of brand names, product names, common names, trade names, product descriptions
etc. even without a particular marking in this works is in no way to be construed to mean that
such names may be regarded as unrestricted in respect of trademark and brand protection
legislation and could thus be used by anyone.

Publisher: Südwestdeutscher Verlag für Hochschulschriften GmbH & Co. KG
Dudweiler Landstr. 99, 66123 Saarbrücken, Germany
Phone +49 681 37 20 271-1, Fax +49 681 37 20 271-0
Email: info@svh-verlag.de

Printed in the U.S.A.
Printed in the U.K. by (see last page)
ISBN: 978-3-8381-2427-8

Copyright © 2011 by the author and Südwestdeutscher Verlag für Hochschulschriften GmbH &
Co. KG and licensors
All rights reserved. Saarbrücken 2011

Contents

	Page
Abbreviations, symbols and units	3
Summary	7
Chapter 1. Background- Interactions between cells and extracellular matrix	9
1.1 The extracellular matrix (ECM)	9
1.1.1 *Composition*	9
1.1.2 *Effects of ECM on cellular functions*	13
1.1.3 *ECM receptors*	14
1.2 Integrins	15
1.2.1 *General aspects*	15
1.2.2 *Integrin structure*	16
1.2.3 *Integrin regulation*	18
1.2.4 *Biological functions of integrins*	23
1.3 Conclusions	24
Chapter 2. Cell adhesion assays- overview	25
2.1 Bulk assays	25
2.1.1 *Techniques using hydrodynamic shear flow*	25
2.1.2 *Centrifugation assay*	27
2.1.3 *Further methods*	28
2.2 Single-cell force spectroscopy (SCFS) techniques	28
2.2.1 *Micropipettes*	29
2.2.2 *Optical tweezers*	30
2.2.3 *Magnetic tweezers*	31
2.2.4 *Conclusions*	31
2.3 AFM force microscopy-based SCFS	33
2.3.1 *Principle- force spectroscopy mode*	33
2.3.2 *AFM-SCFS- experimental setup*	35
2.3.3 *Interpretation of SCFS F-D curves*	38
2.3.4 *AFM-based SCFS- state of the art*	42
2.4 Conclusions	45
Chapter 3. The Bell-Evans model	47
3.1 Basic reaction equations & bond kinetics for receptor-ligand interactions	47
3.2 Dissociation kinetics *near* and *far* from equilibrium	49
3.3 Binding strength	53
3.4 Conclusions	55

Contents

	Page
Chapter 4. Quantifying early steps of $\alpha_2\beta_1$-integrin mediated cell adhesion to collagen type I	**57**
4.1 Abstract	57
4.2 Introduction	58
4.3 Results & Discussion	61
4.3.1 *Preparations*	61
4.3.2 *Investigating single $\alpha_2\beta_1$-integrin mediated adhesion events*	69
4.3.3 *Dependence of $\alpha_2\beta_1$-mediated adhesion on contact time*	74
4.3.4 *Role of actomyosin contractility in cooperative integrin binding*	80
4.3.5 *Visualizing paxillin redistribution during SCFS*	82
4.4 Conclusions & Outlook	84
Chapter 5. Effects of cryptic integrin binding site exposure in collagen type I on osteoblast adhesion and matrix mineralisation	**87**
5.1 Abstract	87
5.2 Introduction	88
5.3 Results & discussion	91
5.3.1 *Characterization of Col and pdCol matrices*	91
5.3.2 *Analysing cellular interactions with Col and pdCol*	93
5.3.3 *Analysing effects of Col and pdCol on cell growth and differentiation*	99
5.4 Conclusions & Outlook	102
Chapter 6. Quantifying adhesion of myeloid progenitors to bone marrow derived stromal cells	**105**
6.1 Abstract	105
6.2 Introduction	106
6.3 Results & discussion	111
6.3.1 *Quantifying overall cell adhesion between 32D and BMSC*	111
6.3.2 *Analyzing single rupture events in F-D curves*	113
6.3.3 *32D cell adhesion to FN and collagen type I coated surfaces*	114
6.3.4 *Role of β_1-integrins in mediating adhesion of 32D cells to BMSC*	116
6.3.5 *β_1-integrin protein and mRNA levels*	117
6.3.6 *32D-V and 32D-BCR/ABL attachment to FN secreted by BMSC*	120
6.4 Conclusions & Outlook	121
Final remarks	**123**
References	**125**
Index of figures and tables	**141**
Appendix (inclusively experimental procedures)	**143**
Acknowledgements	**177**

Abbreviations, symbols and units

Abbreviations

AFM	atomic force microscopy
ABL	Ableson leukemia virus
BCR	break point cluster region
BCR/ABL	fusion protein due to gene fusion of BCR and ABL
BMSC	bone marrow stromal cells
BDM	butandione-2-monoxime
BSA	bovine serum albumin
CAM	cell adhesion molecule
CAM-DR	cell adhesion-mediated drug resistance
CHO	chinese hamster ovary
CML	chronic myeloid leukemia
Col	collagen type I matrix used in this work
DAB	3.3´-diaminobenzidine
DDR	discoidin domain receptor
DFS	dynamic force spectroscopy
DMEM	Dulbeccos modified eagle medium
DNA	deoxyribonucleic acid
ECM	extracellular matrix
EDTA	ethylenediaminetetraacetic acid
EGTA	ethylene glycol-bis(2-aminoethylether)-N,N,N',N'-tetraacetic acid
EM	electron microscopy
FAK	focal adhesion kinase
FCS	fetal calf serum
F-D	force-distance
FITC	fluorescein isothiocyanate
FN	fibronectin
GFOGER	glycine-phenylalanine-hydroxyproline- glycine-glutamate-arginine
HRP	horse radish peroxidase
ICAM	intercellular adhesion molecule

IL-3	interleukin-3
IM	imatinib mesylate
Itgb1	gene encoding β_1-integrin
JAM	junctional adhesion molecule
LFA	lymphocyte function-associated molecule
mAB	monoclonal antibody
MEM	minimal essential medium
MIDAS	metal ion dependent adhesion site
MMP	matrix metallo-protease
MSC	mesenchymal stem cell
pAB	polyclonal antibody
PBS	phosphate buffered solution
pdCol	partially denatured collagen type I matrix used in this work
PFA	paraformaldehyde
RGD	arginine-glycine-aspartic acid
RNA	ribonucleic acid
RPMI	Roswell Park Memorial Institute
RT	room temperature
SCFS	single-cell force spectroscopy
SD	standard deviation
SMFS	single-molecule force spectroscopy
TRITC	tetramethylrhodamine isothiocyanate
VCAM	vascular cell adhesion molecule
VLA	very late antigen

Symbols

a_{eff}	effective cantilever area [μm^2]
A	area [μm^2][m^2]
E	elasticity [Pa]
f, F	force [N][pN][nN]
f^*	mean or most probable rupture force [pN]

F_D	detachment force [pN][nN]
h_{eff}	effective cantilever height [μm]
j	single rupture event
k_B	Boltzman constant ($1.3806504*10^{-23}$ J/K)
k_{eff}	effective spring constant [pN/nm]
k_{off}	rate of bond dissociation [sec^{-1}]
k_{on}	rate of bond formation [$M^{-1}*sec^{-1}$]
L	ligand
η	viscosity [Pa*s]
$N_{L,R}$	number of ligands, receptors
p	pressure [Pa]
P	probability
r_{eff}	effective loading rate [pN/sec]
rpm	rotations per minute [min^{-1}]
R	receptor
τ	lifetime [sec]
t	membrane nanotube unbinding event
v	velocity [μm/sec]
W_D	detachment work [J]

Units

Å	Angstrom (10^{-10} m)
°C	degree Celsius
K	Kelvin
M	Molar (mol/l)
m	meter
μm	mikrometer (10^{-6} m)
N	Newton (kg*m/s^2)
nm	nanometer (10^{-9} m)
nN	nanonewton (10^{-9} N)
pN	piconewton (10^{-12} N)
Pa	Pascal [N/m^2]
sec	seconds

Summary

Interactions of cells with their environment regulate important cellular functions and are required for the organization of cells into tissues and complex organisms. These interactions involve different types of adhesion receptors. Interactions with extracellular matrix (ECM) proteins are mainly mediated by the integrin family of adhesion molecules. Situations in which integrin-ECM interactions are deregulated can cause severe diseases. Thus, the mechanisms underlying integrin-binding and regulation are of high interest, particularly at the molecular level.

How can cell-ECM interactions be studied? While there are several methods to analyze cell adhesion, few provide quantitative data on adhesion forces. One group, single-cell force spectroscopy (SCFS), quantifies adhesion at the single-cell level and can therefore differentiate the adhesive properties of individual cells. One implementation of SCFS is based on atomic force microscopy (AFM); this technique has been employed in the presented work. Advantageously AFM-SCFS combines high temporal and spatial cell manipulation, the ability to measure a large range of adhesion forces and sufficiently high-force resolution to allow the study of single-molecule binding events in the context of a living cell. Since individual adhesion receptors can be analyzed within their physiological environment, AFM-SCFS is a powerful tool to study the mechanisms underlying integrin-regulation.

The presented work is split into six chapters. Chapter one gives background information about cell-ECM interactions. In chapter two, different adhesion assays are compared and contrasted. The theoretical Bell-Evans model which is used to interpret integrin-mediated cell adhesion is discussed in chapter three. Thereafter, the three projects that form the core of the thesis are detailed in chapters four through six.

In the first project (chapter 4), $\alpha_2\beta_1$-integrin mediated cell adhesion to collagen type I, the most abundant structural protein in vertebrates, was quantified using CHO cells. Firstly, $\alpha_2\beta_1$-collagen interactions were investigated at the single-molecule level. Dynamic force spectroscopy permitted calculation of bond specific parameters, such as the bond dissociation rate k_{off} (1.3 ±1.3 sec^{-1}) and the barrier width x_u (2.3 ±0.3 Å). Next, $\alpha_2\beta_1$-integrin mediated cell adhesion to collagen type I was monitored over contact times between 0 and 600 sec. Thereby the kinetics of

$\alpha_2\beta_1$-integrin mediated interactions was explored and insights into the underlying binding mechanisms were gained.

In the second project (chapter five), effects of cryptic integrin binding sites within collagen type I exerted on pre-osteoblasts were investigated. Collagen type I matrices were thermally denatured which lead to exposure of cryptic RGD (Arg-Gly-Asp)-motifs. As a consequence pre-osteoblasts enhanced their adhesion to denatured collagen. Compared to native collagen type I, adhesion to denatured collagen was mediated by a different set of integrins, including α_v- and $\alpha_5\beta_1$-integrins. Cells grown on denatured collagen showed enhanced spreading and motility, which correlated with increased focal adhesion kinase phosphorylation levels. Moreover, osteogenic differentiation kinetics and differentiation potential were increased on denatured collagen. The findings of this project open new perspectives for optimization of tissue engineering substrates.

In the third part (chapter six), the effect of the fusion protein BCR/ABL, a hallmark of chronic myeloid leukemia, on adhesion of myeloid progenitor cells was studied. Adhesion between BCR/ABL transformed progenitor cells to bone marrow derived stromal cells and to different ECM proteins was quantitatively compared to that of control cells. The tyrosine kinase activity of BCR/ABL enhanced cell adhesion, which was blocked by imatinib mesylate, a drug interfering with BCR/ABL activity. BCR/ABL-enhanced adhesion correlated with increased β_1-integrin cell surface concentrations. Since adhesion of leukemic cells to the bone marrow compartment is critical for the development of drug resistance, the reported results may provide a basis for optimized target therapies.

In the three described projects AFM-based SCFS was applied to investigate early steps of integrin-mediated adhesion at the molecular level. Taken together, the results demonstrate that AFM-SCFS is a versatile tool that permits monitoring of cell adhesion from single-molecule interactions to the formation of more complex adhesion sites at the force level.

Chapter 1. Background

Interactions between cells and extracellular matrix

1.1 The extracellular matrix (ECM)

1.1.1 Composition

The extracellular matrix (ECM) represents the authentic substrate for most cells in living organisms[1]. It is a three-dimensional and complex structure composed of collagens, adhesive glycoproteins, proteoglycans and glycosaminoglycans (Table 1). These macromolecules are secreted by cells and locally assembled into an organized network providing a scaffold to embedded cells (Fig. 1)[1-4]. To further increase the complexity of the ECM microenvironment, soluble factors such as growth factors, cytokines, matrix metalloproteinases and other enzymes are present in the ECM.

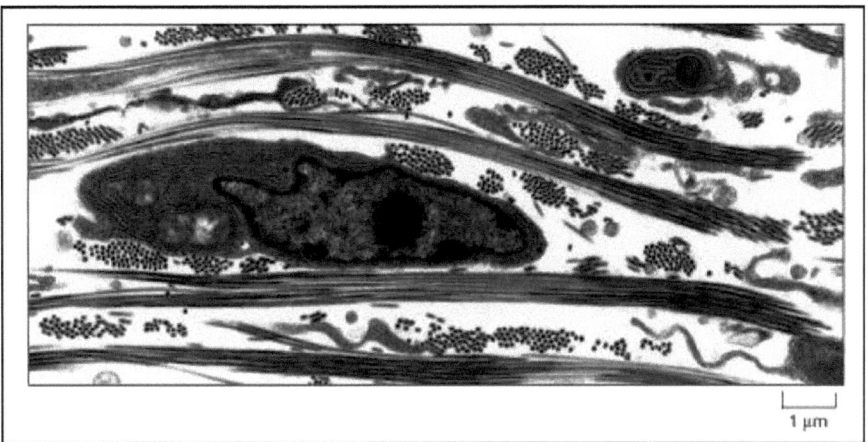

Fig. 1. Chicken fibroblast in the connective tissue of the skin. The cell is surrounded by an ECM composed of thick bundles of collagen type I fibrils (some fibrils are seen as cross-section) (figure taken from[5]).

Composition and density of ECM macromolecules vary significantly in different tissues and organs (Table 1)[6]. For instance, the bone matrix is a composite material consisting of mainly collagen type I fibrils and an inorganic component, hydroxyapatite. Whereas collagen fibrils confer tensile strength, hydroxyapatite makes the bone structure rigid[7, 8]. Collagen type I also significantly contributes to the extremely high tensile strength of tendons[9]. The basal lamina in blood vessels, mainly consisting of laminin and collagen type IV, represents a physical barrier to soluble molecules and a scaffold for aligning cells. Moreover, the ability of the cornea to transmit light while being mechanically resilient can be attributed to an ECM consisting of orthogonal sheets of highly organized collagen fibrils[10]. The given examples show that collagens play an essential role in shaping and strengthening of the ECM in different tissues and organs. Indeed, collagens are the most abundant proteins in vertebrates, contributing to approximately 25 % of their whole protein mass. Apart of fulfilling mechanical functions, collagens further provide a scaffold for the attachment of numerous other proteins, such as adhesive glycoproteins[11]. Subsequently structural features explaining the exceptional mechanical properties of collagens are detailed.

Type of ECM macromolecule	Examples	Predominant localisation
Collagens	Collagen type I (in total >28 different sub-types)	bone, tendon, skin
Adhesive Glycoproteins	Fibronectin	many tissues, increased in wounds
	Laminin	epithelium
	Osteopontin	bone, kidney
	Vitronectin	blood
Proteoglycans	Aggrecan	cartilage
	Decorin	connective tissue
	Perlecan	epithelia, muscle
	Fibromodulin	cartilage, skin, tendon
Glycosaminoglycans	Hyaluran	(cartilage, connective tissues, basement membranes)
	Chondroitin	
	Keratan	
	Heparan Sulfate	

Table 1. Overview about ECM macromolecules. (modified from[6]).

Collagens

There are at least 28 different types of collagen molecules in vertebrates[12]. The basic building block of all collagens is a right-handed triple-helix, formed by three individual polypeptide strands, the so-called α-chains. The polypeptide strains contain approximately 1000 residues and have a length of about 300 nm. The alpha-chains display the repeating structure G-X-Y, in which every third amino acid is glycine, the amino acids X and Y are often proline and hydroxyproline[13]. The polypeptide chains form a left-handed helix that intertwines with two other helices into a right-handed triple-helical structure. This triple-helix can be homotrimeric or heterotrimeric, which is dependent on the collagen type[14]. For instance, Collagen type I is composed of different types of α-chains, two $α_1(I)$- and one $α_2(I)$-chain.

Several collagen types are organized into fibrils (type I, II, III, V, XI). Such collagen fibrils are the most important element providing tensile strength within the ECM in most animal tissues[12, 15]. Other collagens build up networks, such as collagen IV, forming an interlaced network in basement membranes, or are organized to anchoring fibrils (e.g. type VII) or beaded-filaments (e.g. type VI). Further collagen types are classified as transmembrane collagens, endostatin-producing collagens (type XV) or fibril-associated collagens with interrupted helices (FACITs)[12].

There are 11 genes encoding for fibrillar collagens in mammals. In fibrillar collagens the triple-helices assemble into larger units, microfibrils and fibrils[16, 17]. Five triple-helices are believed to form the microfibril, the building block for higher organized fibrils[15, 16]. EM images of negatively stained fibrils and AFM topographs of native fibrils reveal the characteristic, multi-banded structure of the collagen fibrils, which regularly repeats at 67nm[18]. This so-called D-periodicity is explained by the regular staggering of the triple-helices that are laterally aligned (Fig. 2)[15, 19]. This arrangement results in gap and overlap regions and thereby alternating regions of protein density within the fibril[10, 15, 20, 21]. The length of collagen fibrils is undetermined and depends on the tissue type as well as the developmental stage of the tissue; fibrils lengths from μm to mm, and diameters from 12 to >500 nm were found[12]. For example, adult bovine cornea contains uniform collagen fibrils with diameters between 30 and 35 nm and a mean length of 900 μm. In contrast, adult tendon fibrils have diameters of up to 300 nm, their length can reach up to several millimeters[14, 22].

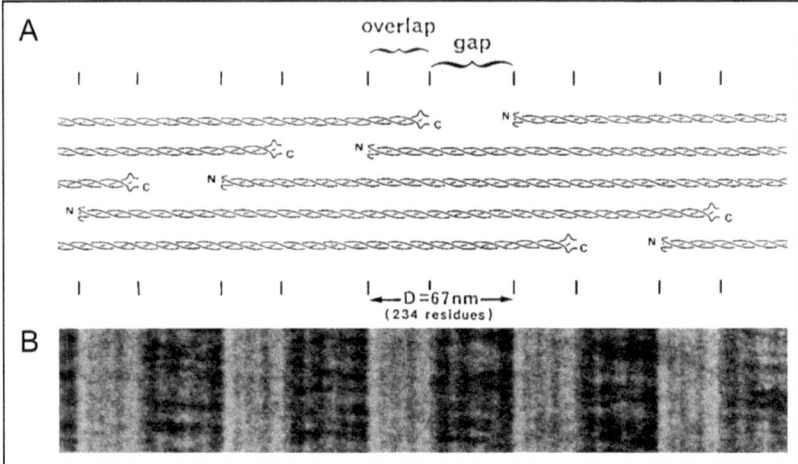

Fig. 2. Axial structure of D-periodic collagen type I fibrils. A. Schematic representation of the axial arrangement of triple-helical collagen molecules within a microfibril. Triple-helices are staggered by D=67nm. B. EM image of negatively stained collagen type I fibrils purified from calfskin (taken from[15]).

Fibril forming collagens are synthesized as procollagens having N- and C- propeptides at both ends of the triple-helical domains. These propeptides are enzymatically cleaved by proteases. The cleaved triple-helices expose telopeptides at their ends, short non-helical extensions. Intact telopeptides are required for collagen fibrillogenesis. Whereas collagen fibrillogenesis can occur *in vitro* as an entropy-driven self-assembly process upon cleavage of the propeptides or from solubilized collagen[23-25], fibrillogenesis *in vivo* is much more complex[26]. It has been proposed that many different proteins participate in this process, for instance fibronectins, integrins and minor collagens serving as organizers and nucleators[26]. However, the exact mechanisms are still not fully understood. Weak noncovalent interactions between the collagen triple-helices assist in fibrillogenesis, but provide only low tensile strength to fibrils[6]. The high tensile strength of collagen fibrils is owed to intra- and inter-molecular covalent crosslinks. These crosslinks are found at residues within the triple-helix and the telopeptides. For most collagens, lysyl oxidases catalyse this crosslinking reaction by activating lysine and hydroxylysine residues[12, 14].

Collagen type I is the most abundant collagen type and the most frequent structural protein in vertebrates. It resides in many tissues and organs, predominantly in skin, bone, tendon and cornea. Aberrant collagen type I synthesis result in a wide spectrum of diseases including *Ehlers Danlos* Syndrome or *osteogenesis imperfecta*[12]. Its predominant occurrence in body tissues together with its excellent biocompatibility and biodegradability make collagen type I the polymer of choice for biomedical and tissue-engineering matrices[27-29]. During the last years, collagen type I has gained the acceptance as a safe material[28]. Common collagen type I sources are collagen rich tissues, such as porcine or cow skin, alternatively recombinant collagen can be used. Collagen type I can be prepared in various formats, as reconstituted soluble collagen or shaped into membrane films, sponges and hydrogels[30]. Medical applications include for instance scaffolds for ligament repair, collagen grafts for scar and burn repair and artificial heart valves. Solubilized collagen or alternatively collagen-like peptides can be used for coating and patterning of non-biological materials to enhance their biocompatibility[27].

1.1.2 Effects of the ECM on cellular functions

Apart of shaping tissues and fulfilling important mechanical functions, the ECM is vital to manifold physiological processes. Cellular interactions with the ECM regulate important cellular processes including cell migration, gene expression, cell survival, tissue organization and differentiation[1, 31-36]. For instance, it is known for a long time that ECM proteins influence the differentiated phenotype of cells[34, 35,37]. It has been demonstrated, for instance, that chondrocytes grown on the ECM protein fibronectin (FN, table 1) adopted a fibroblastic phenotype; this effect could be reversed in absence of FN[37]. Another early study demonstrated that tissue fibroblasts grown in vivo in contact with bone powder transformed into osteoblasts[38]. Furthermore, perturbing ECM-cell interactions during embryonic development leads to severe mutations[35, 39-42]. The role of cell-ECM interactions on cell survival is further demonstrated by the fact that cells that loose contact to their ECM usually undergo apoptosis[43-45]. This mechanism, called anoikis, is crucial for the establishment and maintenance of tissue architecture[46]. There exists also a close interplay between ECM-Cell interactions and growth factor signalling. Growth factors act in concert with ECM molecules and their receptors to promote cell proliferation[47-49]. Growth factor signalling can further activate intracellular signalling pathways that regulate expression of genes encoding for ECM proteins and their receptors[48, 50-53]. In opposite direction cellular interactions with ECM proteins can alter the synthesis of growth factors and respective receptors[48]. Additionally, growth factors can bind directly to ECM proteins and are thereby presented at high local concentration to cells. Some ECM proteins themselves possess mitogenic activity, for instance laminin, tenascin and thrombospondin-1[35].

The mentioned examples show that cells receive specific signals from surrounding ECM proteins that have an influence on essential intracellular signalling pathways. The transfer of signals from the ECM into the cell requires an interface, such as provided by transmembrane ECM binding adhesion molecules.

1.1.3 ECM receptors

By affinity chromatography and experiments using adhesion-blocking antibodies specific membrane glycoproteins that are ECM receptors have been identified[54, 55]. Most of the discovered molecules are part of the integrin family of cell adhesion molecules (CAMs). CAMs are classified into different families, with the main families being cadherins, integrins, selectins and adhesion molecules of the immunoglobulin family. All CAMs are transmembrane proteins, with an ectodomain engaging the ligand, a transmembrane domain and a cytoplasmic domain, which interacts with manifold cytoplasmic proteins. CAM ectodomains are usually large multi-domain structure (20 -50 nm) projecting out from the lipid bilayer. Since the cell surface is relatively rough, this "spacer" distance enables CAMs to bind their ligands[56]. Whereas some CAMs mediate homotypic interactions between cells (e.g. cadherins), others mediate heterotypic interactions with either other cells (selectins, integrins) or ECM proteins (mainly integrins). Other ECM-binding molecules are a diverse group of cell surface proteoglycans. They include among others different lectins, cd44, syndecans, cd36[57] and discoidin domain receptors (DDR)[58]. Some of these bind to collagens (e.g. cd44, cd36, DDR) [58], but also other binding partners within the ECM have been reported[35]. Whereas some of them directly mediate adhesion to ECM proteins, others play a regulatory role in cell adhesion, for instance syndecans[59, 60] and galectins[61, 62]. Due to the dominant role they play in cell-ECM interactions, integrins are introduced subsequently.

1.2 Integrins

1.2.1 General aspects

Integrins are heterodimeric transmembrane glycoproteins found in all metazoans. They are built up by two non-covalently associated α- and β-subunits[63]. So far, 18 α- and 8 β-subunits have been described that can assemble into 24 different heterodimers (Fig. 3). Alternative splicing further enhances the diversity of integrin isoforms. Integrins represent the major adhesion receptors for ECM proteins, few of them also participate in cell-cell adhesion (e.g. $\alpha_4\beta_1$, $\alpha_L\beta_2$).

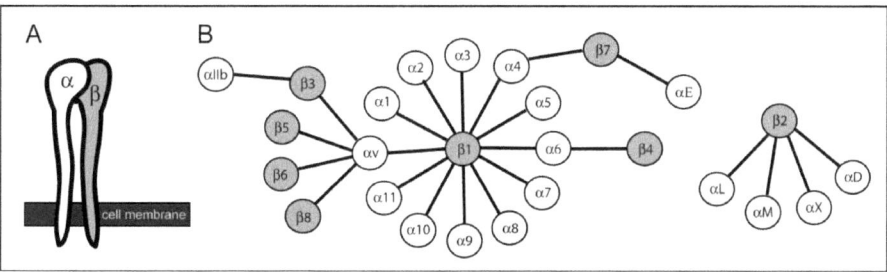

Fig. 3. Overview of integrin heterodimers. (A) Sketch of an integrin heterodimer composed of transmembrane α- and β-subunits. (B) 18 α- and 8 β- subunits can assemble to 24 different types of integrin heterodimers (adapted from [63]).

Integrin-ECM interactions appear to be redundant: several integrin heterodimers can bind to a particular ECM protein (e.g. FN), and most integrins can interact with different ECM proteins (Fig. 4).

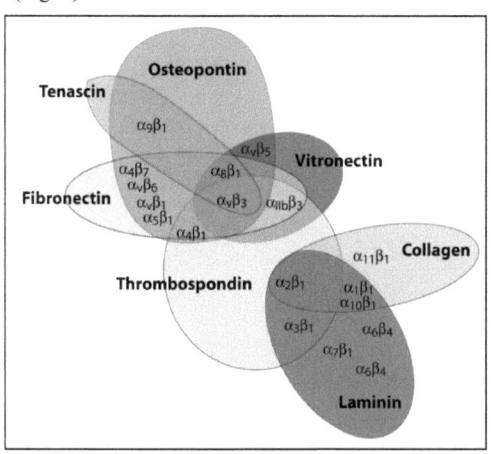

Fig. 4. Combinations of integrin-ECM interactions. Colour- marked circles present different ECM proteins. Binding integrins are inserted. The apparent redun-dance of integrin-ECM interactions is illustrated by the overlapping circles and the multiple receptors within each circle. Note that the shown ECM proteins are only few examples. The figure was created on the basis of integrin ligands reported in. [64].

1.2.2 Integrin structure

The overall shape of the integrin ectodomain is well known from EM images[65, 66]. Integrins have an extracellular globular head (diameter approximately 70 Å), consisting of the N-terminal domains of the α- and β-subunits. The head domain is linked via a long (≈ 100 Å) rigid stalk to a pair of membrane-spanning helices and short cytoplasmic tails*[67, 68]. X-ray crystallography data of integrin ectodomains have predominantly contributed to the understanding of the architecture of integrins. The first integrin crystal structure -the one of integrin $\alpha_v\beta_3$- was obtained in 2001 without bound ligand[67] and later in presence of bound ligand[69]. The N-terminal domains of all α-subunits form a seven-bladed β-propeller that assembles together with the N-terminal domain of the β-subunit (βI-domain) to form the globular head structure (Fig. 5). Glycosylations within the β-propeller further contribute to stabilize the heterodimer[70].

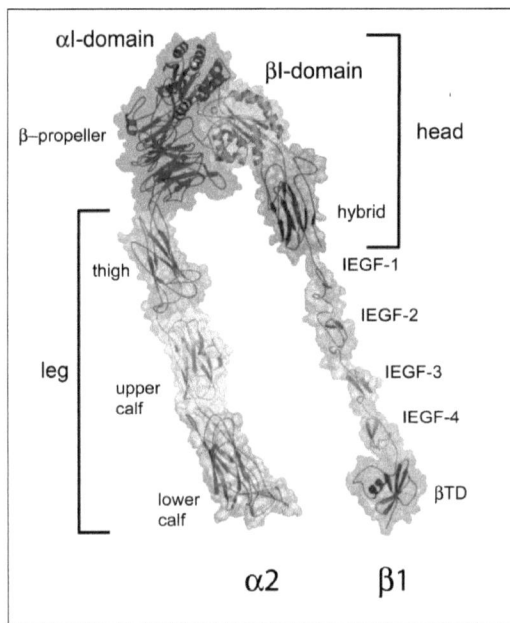

Fig. 5. Molecular model of integrin $\alpha_2\beta_1$ ectodomain in its extended conformation. On the left, the α_2-subunit is shown with its domains. On the right, the β_1-subunit domains are depicted. Divalent metal ions are illustrated as spheres in the αI and βI domains. Collagen binding occurs at the I-domain of the α-subunit (taken from [71]).

---

* β_4-integrin subunit is an exception, it has a long cytoplasmic tail[70].

Two major groups of α-integrins can be distinguished that either contain or lack the so-called "αI-domain" or "von Willebrand factor type A"-domain. αI-domain comprising integrins include all collagen-binding integrins ($α_1β_1$, $α_2β_1$, $α_{10}β_1$, $α_{11}β_1$) and furthermore integrins $α_Dβ_2$, $α_Eβ_2$, $α_Lβ_2$, $α_Mβ_2$, $α_Xβ_2$. Crystal and NMR structures of several integrin αI-domains are available, for instance of $α_2I^{68}$, $α_1I^{72}$, $α_MI^{73}$, $α_LI^{74, 75}$. $α_2I$-domain structures have been resolved in presence and absence of their ligands[68, 76]. The αI-domains are inserted in the β-propeller of the α-subunit, looping out of between blades 2 and 3[68, 70, 71] (Fig. 5). They comprise about 200 amino acids and adopt a so-called Rossmann-fold with a central β-sheet surrounded by α-helices[68](Fig. 6).

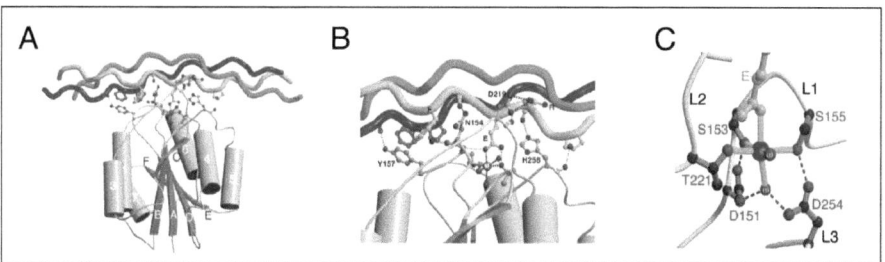

Fig. 6. Structure of the $α_2I$-domain in complex with a collagen like peptide. (A) The shown structure was derived from crystallization experiments using a recombinant $α_2I$-domain and a triple-helical collagen peptide comprising the $α_2$-integrin high-affinity binding motif GFOGER [68]. I-domain helices are drawn as cylinders, β-strands as arrows. (B) Blow-up of (A) revealing details of the interactions between side-chains at the I-domain-ligand interface. Selected side-chains are represented as ball-and-stick. The metal ion is labeled with "M".GFOGER motifs of two collagen strands are involved in binding (middle and trailing strand). Interactions include Van der Waals interactions, hydrogen bonds (dotted lines) and ionic interactions. (C) MIDAS motif of the I-domain. Coordinating side-chains are shown as ball-and-stick. The three loops coordinating the metal are shown. (Figure taken from [68]).

The αI-domains contain a conserved site at which metal ions are bound, the so-called MIDAS site (metal ion dependent adhesion site). It has been shown by mutagenesis studies that the MIDAS motif and surrounding side chains form the contact site with the ligand (Fig. 6)[68]. In the ligand bound state, side-chains forming the MIDAS motif and a negatively charged side residue in the ligand (glutamate in case of collagen $α_2β_1$), coordinate a central metal ion[76, 77] (Fig. 6).

In integrins that do not possess an I-domain, e.g. $α_vβ_3$, the ligand binds at the interface of the α-subunit β-propeller and the βI-domain[67, 68, 73]. Many integrins of this group bind to RGD (arginine-glycine-aspartic acid) motifs within their ligands. It was shown that the aspartic acid of the RGD motif coordinates a metal ion-occupied MIDAS within the βI-domain by a similar mechanism as shown for αI-domains[69, 70].

1.2.3 Integrin regulation

In most biological situations cell adhesion is mediated by multiple adhesive interactions of same or different types of CAMs. The requirements on these adhesive interactions can be quite diverse: in some situations adhesive contacts must be quickly assembled and disassembled (e.g. during cell migration). Other adhesive contacts need to be strong to resist mechanical stress (e.g. in contracting muscle). Thus, adhesive interactions have to be precisely regulated not only quantitatively, but also locally and temporally.

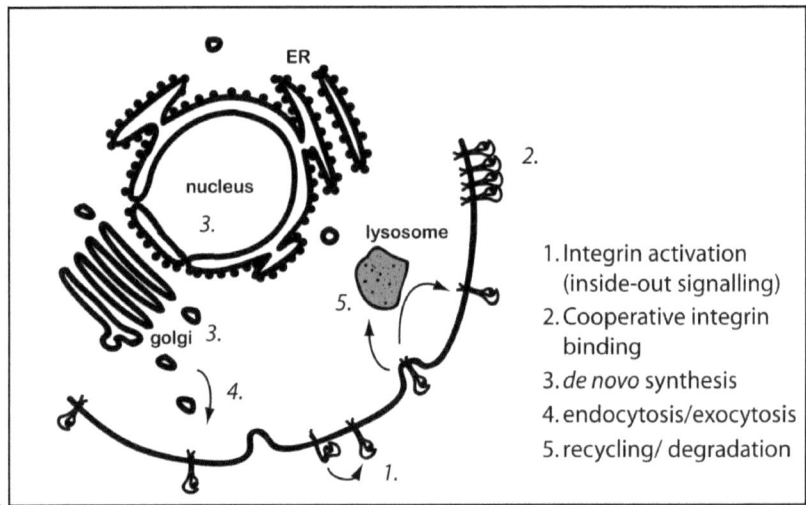

Fig. 7. Different mechanisms to regulate cell surface concentrations of integrins. 1. Altering the binding affinity of individual receptors by inside-out signalling[78]. 2. Controlling cooperative receptor binding[63, 77, 79]. 3. Regulating de novo protein synthesis, for instance at transcriptional and translational level. 4. Transport of CAMs to the cell surface, 4. Receptor endocytosis and recycling or degradation[80, 81]. Taken with modifications from [81].

Fig. 7 summarizes different strategies that can be used by cells to regulate adhesion, the focus is thereby on integrin-mediated adhesion. Firstly, avidity of integrin binding, the overall adhesion, can be modulated. Avidity regulation includes both, regulation of the affinity of individual integrins for their ligands (activation) (Fig. 7-1), but also cooperative integrin binding at sites of clustered integrins (Fig. 7-2). Moreover, total integrin concentrations can be altered by controlling integrin *de novo* synthesis and maturation, degradation and recycling of integrins. Last, cell surface concentrations can be tuned by controlled endocytosis and exocytosis[81].

In the following mechanisms underlying integrin activation and integrin clustering will be further explained.

Integrin activation- inside out signalling

Two principal quaternary conformations have been described for integrins, an extended and a bent one[71, 77] (Fig. 8). These have been attributed to the active/high-affinity and the inactive/low affinity states of the integrin[82]. The conversion of the bent into the extended integrin conformation and increase in ligand-binding affinity (here also called integrin activation) is regulated by intracellular signals, in a process called *inside-out signalling*. In the last years, important insights into the mechanisms underlying integrin activation have been obtained. Mutagenesis studies showed that the membrane-proximal regions of the cytoplasmic tails play a pivotal role in integrin activation[83]. The integrin is kept in its low-affinity state when the cytoplasmic tails are bound together[71, 84, 85]. This association is stabilized by a salt bridge between the membrane-proximal regions of the cytoplasmic integrin α- and β-tail (Fig. 8). Upon *inside-out signalling* the salt bridge between α/β-tails is disrupted and the tails are separated. This leads to major conformational changes, resulting in an extended integrin conformation[84, 86]. Since the membrane proximal regions are well conserved within α- and β-subunits, similar mechanisms might be responsible for the activation of different integrin heterodimers.

It has been shown that *inside-out* activation of β_1- and β_3-integrins requires talin binding to regions far from the membrane-proximal region[83, 87, 88]. In addition, binding of cytohesin-1[89] and β3-endonexin[90] were shown to result in β_2- and β_3-integrin activation[88].

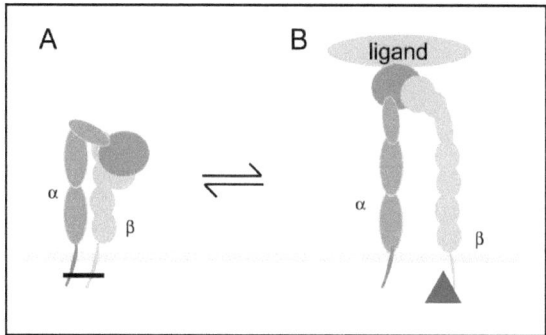

Fig. 8. Integrin structural changes during activation. Integrin in its inactive, low-affinity (A) and active, high-affinity conformation (B).

Many different mechanisms were proposed by which talin-binding to integrin tails might be regulated. These include integrin phosphorylation, talin proteolysis, talin activation by phosphatidylinositol (4,5)-bisphosphate and competition between integrin tail-binding proteins[83]. Recently it has been shown that kindlin-2 is required for talin-induced integrin activation[91]. However, so far the exact mechanisms have not been elucidated. Whereas the integrin β-tail is involved in regulating integrin activation upon binding to signalling proteins, the α-tail appears to contribute to the cell-type specificity of integrin activation[92].

Since the initial events leading to integrin activation occur at the membrane-proximal regions of the cytoplasmic integrin tails, long-range conformational rearrangements must propagate through the transmembrane and stalk regions up to the ligand-binding site[77, 93]. Crystal structure data of I-domains revealed structural variations between ligated and unligated states which were attributed to high-affinity/closed and low-affinity/open conformations[76, 94]. The interpretation of these structural data has been highly controversial in the integrin field[77]. The critical question was if the conformational changes within the I-domain were physiologically relevant for affinity-regulation of the I-domain, or if they resulted from induced fit upon ligand binding[77]. The so-called *switchblade* model postulates that integrin leg separation and extension are directly coupled to conformational changes within the ligand binding site. Thus, integrin extension switches the ligand-binding site from its low-affinity to its high-affinity state[95]. In accordance with the *switchblade* model, there are two effects of integrin extension that contribute to high-affinity binding: in the bent integrin conformation, the ligand-binding head domains are in proximity to the membrane and sterically hinder ligand binding. Integrin leg extension lifts the ligand-binding site that is faced toward the membrane in the low-affinity state up and orients it pointing away from the cell membrane towards a potential ligand. Thereby the ligand-binding domain within the head is exposed to the ECM and better accessible for ligands[71, 82]. Secondly, the conformational changes within the I domain occurring as a consequence of leg extension additionally increase its affinity for its ligand. Thus, following the switchblade model, extended integrins represent the high-affinity, bent integrins the low-affinity conformation [71, 77, 82].

Another model, the so-called *deadbolt* model provides an alternative interpretation. According to this model, also integrins in bent conformation may bind their ligands, which as a consequence leads to integrin extension. The deadbolt model was supported by EM images that found soluble $\alpha_v\beta_3$ bound to FN fragments in its bent conformation. However, these findings might be explained by fact that truncated and soluble integrins were used for analysis and the use of small fragments instead of the entire FN molecule. Recent experiments clearly favour the *switchblade* model[82].

Outside-in signalling

Integrins that are not bound to their ECM ligands are supposed to be diffusely distributed over the cell surface and not linked to the actin cytoskeleton[96]. Binding to an ECM ligand stabilizes the integrin in its extended, high-affinity conformation[77]. In the high-affinity conformation the cytoplasmic integrin tails are separated and cytoplasmic components can bind[71, 77, 82, 85]. Thus, ligand binding facilitates association of integrin tails with the cytoskeleton[97, 98].

Cytoskeleton-associated integrins can form clusters that contribute to enhanced cell adhesion by increasing integrin avidity for its ligands [96, 99, 100]. It has been speculated that integrin clustering is initiated by the multivalent nature of ECM proteins: ligand binding stabilizes integrins in a certain spatial arrangement that supports cluster formation[77, 101]. Controversially, it has been proposed that integrin clustering occurs upon signals from inside the cells. Rho A activation and increased acto-myosin contractility have been implicated in this process[102-104]. Possibly these mechanisms are not mutually exclusive. Integrin-clustering does not only contribute to enhanced mechanical anchorage of cells to ECM proteins, but also has an important role in regulating important cellular functions. Upon integrin-clustering signalling proteins and cytoskeletal proteins are recruited to the cytoplasmic sites and thereby intracellular signalling pathways controlling cell differentiation[81, 105, 106], proliferation[107] and survival[108, 109] are triggered (Fig. 9). Since these pathways are initiated by an extracellular stimulus (=ligand binding) this process is referred to as *outside-in signalling*.

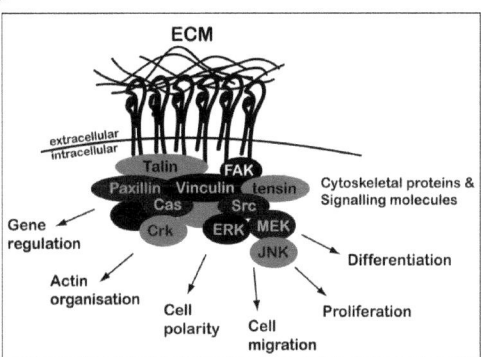

*Fig. 9. **Integrin-cytoskeleton interactions.** Integrin clusters associate with multiple intracellular proteins and thereby trigger signalling pathway regulating important cellular processes.*

In Fig. 10 the mechanisms of *inside-out* and *outside-in* signalling are summarized. It is underlined that the shown events are not sequentially following mechanisms, but that there is a dynamic change between the different states (bend, extended, cytoskeleton-associated, clustered). The equilibrium between the respective states can be shifted upon binding of intracellular or extracellular binding partners.

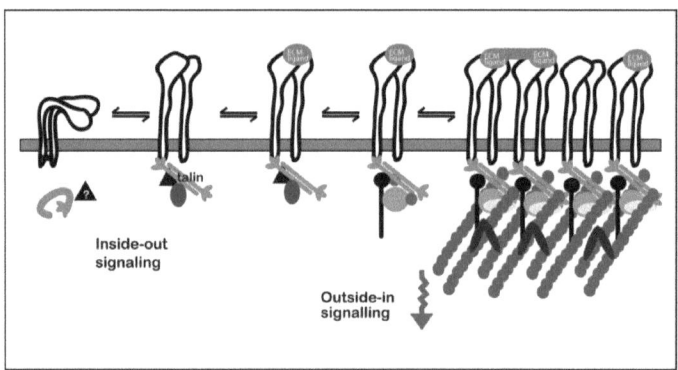

Fig. 10. Integrin inside-out and outside-in signalling. In the inactive integrin conformation (bent) interactions between cytoplasmic tails are stabilized by a salt bridge (black bar). Talin binding induces integrin activation (extended conformation) (inside-out signalling). In the active conformation integrins can bind to extracellular ligands. Ligand binding stabilizes the active conformation with separated cytoplasmic tails and intracellular ligands can bind to tails. Integrin clustering is regulated by multivalent extracellular ligand binding and/or intracellular events involving acto-myosin contractility. Association with intracellular signalling proteins activates intracellular signalling pathways (outside-in signalling).

Higher order adhesion sites

Above mentioned integrin clusters can be visualized by fluorescence microscopy as dynamic, dot-like structures at the edges of lamellipodia[110-113]. They are called focal complexes. Focal complexes are believed to be precursors of focal adhesions* (also termed focal contacts) at which strong cell-matrix adhesion occurs (Fig. 11)[115]. Focal adhesions are sites of high protein density that were identified long ago using interference-reflections microscopy and electron microscopy [116, 117]. In focal adhesions, a tight contact exists between membrane and substrate, leaving a gap of only 10-15 nm[117, 118]. Maturation of focal complexes into focal adhesions is

---

* Beside focal adhesions, there exist also other types of matrix adhesions, such as fibrillar adhesions and podosomes, but these will not be discussed here[114, 115].

generally accepted to be a consequence of Rho activation or external force application[102, 104, 110, 112, 119].

The major transmembrane ECM receptors within adhesion sites are integrins, but also others, such as preoteoglycans[120], glycosaminoglycan receptors such as syndecans[121] and signalling proteins are present[122-124]. Moreover, >50 cytoplasmic proteins localize to focal adhesions, e.g. cytoskeletal proteins (e.g. vinculin, α-actinin, talin), tyrosine kinases (e.g. Src, FAK), serine /threonine kinases (e.g. ILK, PAK), modulators of small GTPases (ASAP1, Graf), tyrosine phosphatases (e.g. SHP-2) and other enzymes (e.g. PI3K, calpain II). Some of these components bind to actin filaments and/or to the cytoplasmic tails of integrins. It has been suggested that the list of focal contact molecules is far from being complete[118].

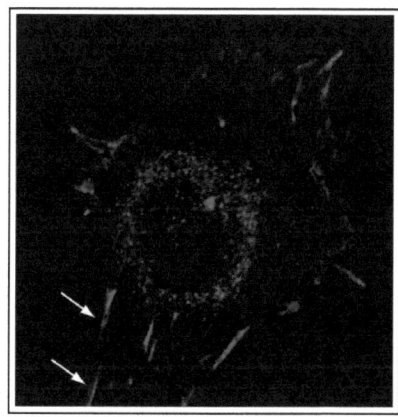

Fig. 11. Focal adhesions visualized by fluorescence microscopy. Mef cells seeded on glass coverslips were fixed, permeabilized and stained for the focal adhesion protein paxillin. Arrows point to two exemplary focal adhesions.

1.2.4 Biological function of integrins

Characterization of integrin knockout mice have revealed important physiological roles of integrins. Although the binding specificities of several integrins are redundant, loss of almost any α- or β-subunits results in biological defects in respective knockout-mice. Such defects range from small imperfections (e.g. α_1-knockout mice[125]) to severe malfunctions that are lethal at certain embryonic stages or shortly after birth (e.g. β_1-integrin[126-128]). Integrins are involved in many physiological processes, for instance in the immune system during inflammations, during wound healing, blood clot formation, during fertilization or embryonic development[129]. Consequently several human diseases correlate with defects of certain integrin encoding genes, for instance:

- Defects in platelet integrins (α_{IIb}- or β_3-subunits) cause the bleeding disorder *Glanzmann's thrombasthenia*[130].
- Defects in the collagen binding integrin α_1-integrin can lead to *fibrosis*[125].
- Mutations of β_2-integrins can result in *LAD* (leukocyte adhesion deficiency)[130].
- Due to their role in cell signalling, angiogenesis and migration and interactions with urokinase receptor, integrins have been also implicated in tumour progression[131, 132].
- Some integrins are targets of viruses (Hantavirus, Papillomavirus, Echovirus and other pathogens, such as *Borrelia burgdorferi*, *Bordetella pertussis* and *Yersinia spp.*[129, 133, 134].
- Agents that are secreted by leeches and disintegrins inherent to toxins e.g. snake venom *jararhagin*[133] are integrin ligands and act as inhibitors of platelet aggregation[129].

Consequently, integrins have been recognized as important targets for drug development. It has been proven in independent studies that antagonists against integrin $\alpha_{IIb}\beta_3$, such as monoclonal antibodies or cyclic peptides, can be used to counteract blood clot formation[79]. Abciximab, a monoclonal antibody, is already admitted for use and indicated for application during and after coronary artery procedures, for instance.

1.3 Conclusions

In the previous sections the importance of integrin-ECM interactions has been discussed. Since these interactions play an essential role in many physiological, but also pathological situations, the basic mechanisms underlying integrin binding and regulation are of great interest. For instance, a better understanding of integrin-ECM may bring benefits to pharmaceutics, since it may aid the development of specific drugs. Moreover, understanding cell-ECM interactions is of particular importance for the design of implant materials that promote cell attachment and growth. How can these interactions be investigated? In the following different techniques that were developed to analyse cell adhesion are compared and contrasted.

Chapter 2

Cell adhesion assays- Overview

There exist many cell adhesion assays that have been developed to qualitatively or quantitatively study cell adhesion. These techniques have in common that cells are allowed to attach for a certain period to an adhesive substrate, for instance a surface coated with purified proteins. Thereafter, cell detachment is induced by a certain method. Common assays employ centrifugal forces, hydrodynamic shear forces or locally applied pulling forces to cause cell detachment. Adhesion assays can be classified by many ways, one is by the number of cells that is analysed in each experiment, either many (*bulk*) or only single cells *(single-cell)* assays). Subsequently respective examples are presented. In Table 2 a systematic overview is provided, and main characteristics, advantages and disadvantages of the techniques are given.

2.1 Bulk assays

2.1.1 Techniques using hydrodynamic shear flow

In hydrodynamic flow assays, a flow of physiological buffer is generated on a surface with attached cells. The cells are thereby subjected to shear forces that eventually cause cell detachment. Depending on the apparatus and the flow configuration, constant flow rates (e.g. parallel plate) or gradients can be applied (e.g. spinning disc). In difference to other adhesion assays, detachment forces generated in hydrodynamic flow systems peel off the cells, since forces act approximately parallel to the substrate (table 2). Disadvantageously, these hydrodynamic shear forces depend on cell geometry and are not uniformly distributed along the cell surface. Therefore, to get an estimate of the applied detachment forces, simplified assumptions have to be used [135].

Washing assays

The simplest and most commonly applied adhesion assays are so-called washing assays. Cells are allowed to attach for a certain time to a substrate of interest, being either a protein-coated surface or to a monolayer of cells. Then the substrate is rinsed several times with physiological buffers and thereby non-attached or weakly attached cells are dislodged from the substrate[136]. For quantification the number of cells that remained attached to the substrate is determined. Cell numbers are either counted, otherwise more convenient methods such as radiolabelling, staining with colorimetric or fluorescent reagents or enzymatic techniques may be used[137]. A major advantage of the washing assays is their simplicity, and the fact that only routine laboratory equipment is needed. It is remarkable that these washing assays could identify adhesion molecules and regulatory mechanisms[135]. The same simplicity, however, carries severe disadvantages as low reproducibility and poor sensitivity of the measurements. First of all, no quantitative data on adhesion forces are obtained. Moreover, the applied shear forces are unknown, unevenly distributed and furthermore difficult to control. Sensitivity is low, consequently small differences in adhesion among different experimental groups that might be of biological relevance are difficult to detect. Whereas short contact times (< 30 min) are difficult to control, the applied forces may not be sufficiently high to detach cells after prolonged contact times, especially when cell are spread on the substrate; this effect can mask eventual adhesion differences of distinct cell populations. Taken together, these limitations might explain the inconsistent results that are frequently obtained among different research groups. However, despite their limitations, washing assays can be a useful tool for certain biological and medical questions, being economic and fast tests to perform. Washing assays were also carried out in the projects presented in chapters five and six.

Spinning disc device

Better-controlled flow conditions are generated using spinning disc devices. Adhesive substrates with attached cells are mounted onto a rotating device and spun in a fluid-filled chamber. Upon rotation of the disc, fluid from the surrounding is axially drawn to the disc center and displaced radially.[135] At laminar flow conditions thickness and radial velocity of the fluid layer carried with the disc are constant. By the liquid flow, well-characterized hydrodynamic forces are applied to the cells; these tangentially acting forces increase linearly with the disc radius, such that negligible hydrodynamic forces act at disc center and maximal forces at the periphery. After spinning, numbers of attached cells per radial position are evaluated. This can be done by staining fixed cells with a DNA-binding fluorescent dye, followed by automated analysis of the fluorescence intensity per radial position[135]. Commonly the shear force, at which 50 % of cells remained attached, is calculated and defined as mean adhesive force. However, the actual force acting on the cells cannot be precisely assessed. Moreover, obtained results critically depend on the cell shape (e.g. spreading morphology) and size. The method is not suitable to perform cell-cell adhesion measurements. Furthermore, initial cell adhesion cannot be investigated, because cell attachment periods < 30 min are difficult to control. Positive aspects of spinning discs devices are that the experiments are well reproducible and a large number of cells can be easily analysed.

Parallel plate chamber

Parallel plate chambers can be used to study both static and dynamic cell adhesion. To perform a static adhesion assay, an adhesive substrate with attached cells is mounted into the flow chamber. Then a well controlled, laminar flow of physiological buffer is pumped through the chamber. Depending on the flow rate, cells may become dislodged. For analysis the fraction of cells that remained attached to the substrate is detected as a function of the applied flow rate. Since a constant flow is produced in this system, several experiments must be conducted at different flow rates to characterize the dependence of attached cell number on the applied shear force.

In dynamic adhesion tests cell attachment to an adhesive substrate is studied in a well defined shear field. Therefore a cell suspension is perfused through the flow chamber that carries an adhesive substrate at its bottom. Since the chamber is mounted on a light microscope equipped with a camera, the movement of single cells can be monitored and recorded. Transient adhesive interactions between cell and substrate cause cell rolling, similar as occurring *in vivo* for blood cells in the blood vessels[138]. Thus, flow chambers are of particular interest to study adhesion events that occur *in vivo* under such flow conditions. Typically frequency and duration of cell arrest on the adhesive substrate are analysed and by applying varying flow rates, bond dissociation rates can be determined[139-142]. However, the effective force acting on the established bonds is unknown. Usually it is assumed that the generated forces are too low (5-10 pN) to alter the natural bond lifetimes. This may be the case if flow chambers are operated at very low shear rates ($1 - 10$ s^{-1})[140, 143]. However, such shear rates are approximately an order of magnitude lower than shear rates found in the blood stream, and therefore not comparable to physiological conditions. Advantageously, flow chambers provide relatively simple and inexpensive systems. Moreover, there are commercially available systems.

2.1.2 Centrifugation assay

In centrifugation assays cells attached to an adhesive substrate experience centrifugal forces that act- in difference to above described hydrodynamic flow assays- vertical to the adhesive substrate (Table 2). Typically microtiter wells are coated with purified proteins of interest. Then, cells are added to the wells and by mounting another microtiter plate on the first one, small chambers of two opposed wells are created. By short centrifugation the cells are gently brought in contact with the coated surface. Then, cells are allowed to attach for a certain time to the wells, until wells are inverted and another centrifugation step is performed. Thereby weakly attached and non-attached cells are released and transferred into the opposite well. Quantification of cells that remained attached to the substrate can be done automatically, e.g. by radio-labelling, staining with a colorimetric or fluorescent reagent or enzymatic techniques. Since only a single force is exerted in each experiment, the experiment has to be repeated at different centrifugation speeds to obtain information about the involved adhesion forces[144, 145].

2.1.3 Further methods

There are also functional assays, as migration or spreading assays that are employed to indirectly study cell adhesion[146]. However, cell spreading or migration do not always directly correlate with adhesion. Although cell adhesion is required for cell spreading, the mechanisms regulating cell spreading are far more complex. Similarly, cell migration is a tightly regulated process. For instance, it was shown for epithelial cells on FN-coated surfaces that there is not a simple linear correlation between adhesion and migration speed[147]. Thus, no reliable information on cell adhesion forces can be provided by such assays. In another approach, focal adhesions are quantified in terms of size of numbers to get insights into the adhesion strength of cells[148-150]. However, no information about adhesion forces can be obtained.

2.2 Single-cell force spectroscopy (SCFS) techniques

In above described *bulk* assays a larger number of cells is tested in each experiment. This has the advantage that easily statistically relevant data can be obtained. However, as discussed above, bulk assays cannot give exact quantitative data on adhesion forces. Moreover, since the average behaviour of a large cell population is analysed, potential differences in the adhesion of individual cells cannot be detected. Such adhesive subpopulations might appear due to different functional states of individual cells. Thus, valuable information might get lost in bulk assays.

For a more quantitative approach, techniques are needed that quantify the adhesion of single cells. Such techniques have been developed within the last years; they are termed single-cell force spectroscopy (SCFS) techniques. SCFS experiments are usually more time-consuming compared to bulk assays, because only one cell is analysed in each experiment. Since adhesive properties of individual cells can vary considerably, even for cells of the same cell line, the experiment has to be repeated many times.

Some SCFS techniques allow cell adhesion to be characterised at the single-molecule level. Respective examples are given below. SCFS experiments with single-molecule resolution permit to identify a certain set of adhesion receptors contributing to overall cell adhesion. Furthermore, single-molecule studies can give detailed insights into regulation mechanisms of adhesion receptors. This is of particular interest for the presented work, in which integrin-mediated adhesion is studied. As discussed in the previous chapter, integrin regulation can involve multiple mechanisms (1.2.3) that might be distinguished by SCFS.

SCFS single-molecule experiments can be further employed to determine biophysically relevant parameters, such as bond dissociation rates (see 3.3). Traditionally binding rates of adhesion receptors are studied using *in vitro* bulk assays with purified molecules, e.g. solid phase assays[151] or batch experiments with radio-labeled ligands[152]. Since in such assays many molecules are examined in parallel, the results reflect the ensemble average of the behaviour of individual molecules. Thus, they are not suitable to reveal different functional states of individual adhesion receptors. Further limitations of traditional binding assays arise from the use of purified adhesion receptors. Separated from their native environment, the membranes of living cells, receptor-binding characteristics might be significantly altered [143, 153]. For instance, the purification process might affect the functionality of the adhesion receptor. Moreover, the spatial orientation of adhesion receptors inserted in the cell membrane is optimized for ligand binding. In contrast, ligand binding affinities of isolated receptors are possibly altered. Moreover, *in vitro* assays are usually restricted to the study of receptor ectodomains or even smaller fragments; their binding characteristics might not be comparable to intact adhesion receptors. A further important issue is, that multiple different intracellular and transmembrane binding partners regulate adhesion receptor binding *in vivo*. Since these are absent in *in vitro* assays, mechanisms that regulate adhesion receptor binding cannot be studied. The given arguments underline that a far better approach consists in analyzing adhesion receptor-ligand interactions in the context of a living cell. In the following several techniques that can be used for SCFS are briefly described.

2.2.1 Micropipettes

A cell is aspirated into a micropipette and allowed to interact with an adhesive substrate. Thereafter, the micropipette is retracted, which leads- depending on the suction pressure- either to cell-substrate or cell-pipette detachment. In case of cell-pipette detachment, the suction pressure in the micropipette- and thereby the force holding the cell (F=p/A)- is stepwise increased, until the cell detaches from the substrate. By averaging the forces applied in the steps prior and at cell-substrate detachment, the adhesion of the cell to the substrate can be approximated. In the described setup, the force resolution is limited to large forces of several 100´s of pN´s and single-molecule binding cannot be assessed.

Micropipettes can also be used in a modified setup to perform single-molecule experiments at a high force resolution (<0.1 pN); this technique is referred to as biomembrane force probe (BFP)[154]. Beside single-molecule interactions with isolated adhesion molecules, this system can also be used to detect cell-protein or cell-cell interactions[155]. To study cell-ECM interactions, a microsphere with immobilized ligand molecules is coupled to the force sensor. This sensor is typically a swollen red blood cell or a phospholipid vesicle, aspirated by a micropipette. Then a cell, held by another pipette, is brought into contact with the ligand-coated microsphere and thereafter separated again. Implemented piezoelements permit precise relative movements of the pipettes. The position of the microsphere attached to the force sensor is monitored via a camera. Its displacement is proportional to the force acting on the microsphere[154, 156, 157]. Since the range of detectable forces is limited to <1 nN[158], BFP is confined to the detection of lower adhesion forces.

2.2.2 Optical tweezers

An object, for instance a microsphere modified with a molecule of interest, is trapped within the focal point of a strongly focused laser beam. The force holding the object in focus of the laser depends on the dielectric properties of the object and the gradient intensity[159]. To measure adhesive forces between the trapped object and a living cell, for instance, the object is brought in contact with the cell and withdrawn. If the object adheres to the cell, the object is displaced from the focal point of the trap. This displacement, detected for instance by differential interference contrast (DIC), can be correlated to the acting force[159]. In such experiments forces within a limited range of 0.1 - 100 pN can be generated and detected[160]. Consequently only single molecule binding can be assessed. Apart of the limited force range, optical tweezers have further limitations: Optical tweezers are very sensitive to optical perturbations. Moreover, the selectivity is limited, since many other dielectric particles near the focus of the lasers will be trapped.

Also entire cells can be trapped and manipulated, for instance bacteria[161] or mammalian cells[159, 162]. Thereby interactions between trapped cells and a substrate can be analysed. However, the possible range of detachment forces is even more limited for trapped cells. The maximal forces that could be measured between mammalian cells and substrates were lower than 50 pN, which is not sufficient to break multiple adhesive bonds[159]. Moreover, the high intensity at the focus of the trapping laser beam results in local heating, which may cause damage to living cells[160]. Although there are commercially available systems, these are often not fully suitable to perform high-resolution measurements, therefore adapting such a system might still be challenging regarding technical knowledge and time.

2.2.3 Magnetic tweezers

The principle of magnetic tweezers consists in the application of a force on a magnetic microsphere within a magnetic field. This technique can be used to investigate cell-protein interactions. Therefore a microsphere is decorated with ligand molecules and brought into contact with a living cell. Thereafter a force is applied onto the bead by generating a magnetic field. In systems with implemented feedback controlled forces up to several nN can be applied, which allows also to study the rupture of more complex cell adhesion sites[163]. However, the actual force acting on the cell cannot be precisely measured. Thus, precise single-molecules measurements are not possible. Furthermore, the position at which the microsphere is in contact with the cell cannot be controlled. The mentioned experimental setup is not suitable to investigate adhesion events at early time-points. Therefore an alternative setup can be used: a magnetic bead is linked to a cell, and the bead-cell-couple is brought in contact with a flat substrate. After a certain contact time, the bead-cell-couple is detached by pulling on the microsphere[164]. Advantageously, the setups can be combined with fluorescence microscopy, and thereby interesting biological questions can be addressed. As an example, magnetic tweezers were successfully used to study the remodeling of focal adhesions under applied mechanical stress[165].

2.2.4 Conclusion

It has been shown that quite a few different techniques exist that study cell adhesion. These techniques differ significantly in their principles and abilities, as summarized in Table 2. Which of the technique is applied for a certain task, clearly depends on the context. First aim of the presented work is the analysis of $\alpha_2\beta_1$-integrin collagen type I interactions at the molecular level (chapter four). This requires a technique with single-molecules resolution, which reduces the presented techniques to BFP and optical tweezers. Furthermore, the kinetics of integrin-mediated adhesion is going to be studied over a time range of 10 min. Since after several minutes cells may establish strong contact to the substrate, a large force range of several nN must be assessable to detach the cell. Thus, BFP and optical tweezers were not suitable for such task. Furthermore the time for which the cell is kept in contact with the adhesive substrates must be precisely controlled to study kinetics. Thus, a technique is needed that combines the mentioned aspects of covering a large force range, detecting adhesion at high force resolution and permitting high temporal and spatial precision of cell manipulation. This can be done by a SCFS technique based on atomic force microscopy (AFM) that is introduced in the following.

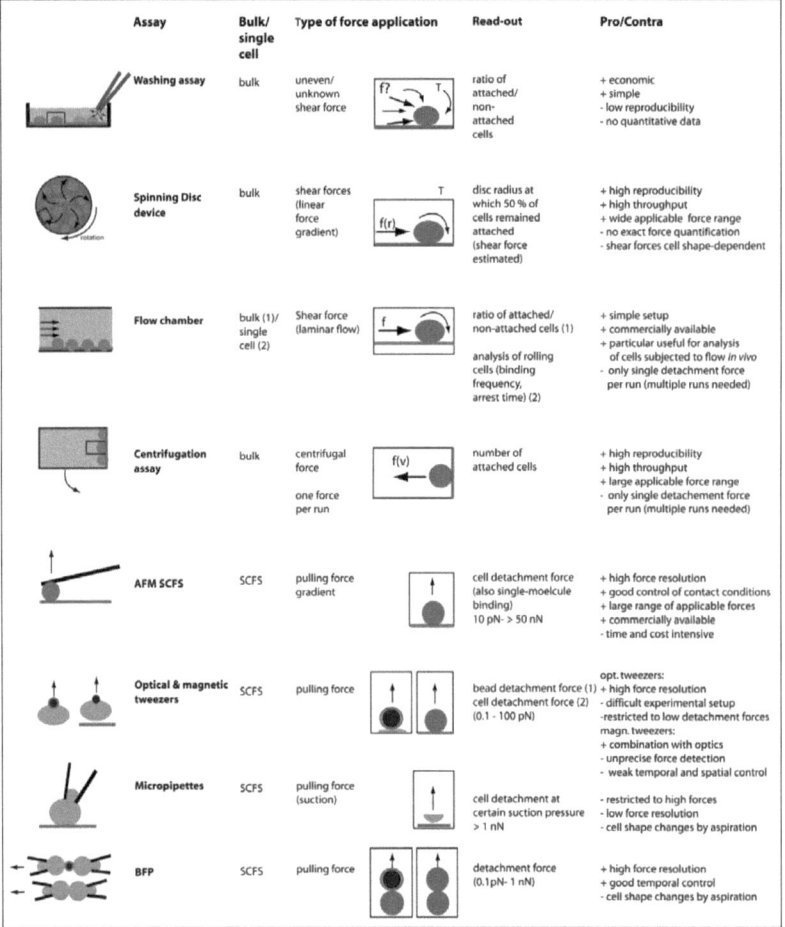

Table 2. Overview about different cell adhesion assays.

2.3 Atomic force microscopy- based SCFS

2.3.1 Principle – force spectroscopy mode

Atomic force microscopy (AFM)[166] has been developed as a tool to analyse the topography of flat samples at sub-nanometer resolution[1]. Its basic principle consists in the detection of ultrasmall forces[166]. The so-called AFM cantilever, a highly flexible spring, represents the key component of the AFM (Fig. 12 A). Forces acting on the cantilever cause it to bend. This bending is - within a limited range- directly proportional to the applied force, according to Hook's law ($F = k \cdot x$). Most commonly cantilever bending is detected via an optical lever method (Fig. 12 B). A laser beam is reflected by the back of the cantilever onto a segmented photodiode[2][170-172]. When the cantilever bends, the position at which the laser beam hits the photodiode changes and different voltage signals are detected in the individual segments of the photodiode. To convert the detected voltage change into a force, the system has to be calibrated prior to experiments (see B1.1).

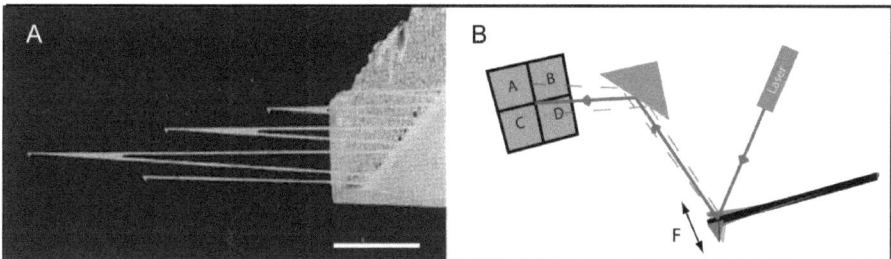

Fig. 12. AFM cantilever and principle of force detection. A. EM image of a chip with five AFM cantilevers. Scale bar corresponds to ~2 μm (taken from www.veeco.com). B. The cantilever bends proportional to the force acting on it. Cantilever bending causes laser beam deflections and as consequence altered voltage signals are detected in the segmented photodiode.

Whereas AFM has been applied as an imaging tool for the last decades, it has been increasingly used during the last years to measure molecular forces. This technique is referred to as *force spectroscopy*. At normal conditions and in liquid environment, AFM force spectroscopy allows the detection of forces down to 10 pN. This permits the detection of weak non-covalent ligand-receptor interactions which break in the pN range[158, 173]. Since AFM force spectroscopy

---

[1] Since AFM imaging is not scope of this thesis, it will not be expanded here.

[2] There exist also alternative detection methods for the cantilever deflection, for instance, interferometric detection[167, 168] or piezo-resistive detection[169].

experiments can be conducted under physiological conditions, functional biological samples can be studied.

In AFM force spectroscopy experiments the X-Y-position of the AFM cantilever is maintained constant, whereas its Z-position is changed in a controlled fashion by piezoelectric elements. During cantilever movement, forces acting on the cantilever are monitored over the covered distance. Thereby a force-distance (F-D) curve is generated. There are several possible variants of force spectroscopy experiments. To specifically test ligand-receptor interactions, the AFM tip is decorated with certain receptor molecules and brought in contact with a ligand-coated surface (Fig. 13 B). When the cantilever is withdrawn again, potential interactions established between tip and substrate bend the cantilever towards the surface (Fig. 13 C) until the tip is released (Fig. 13 D). From the F-D retraction curve the interaction force between tip and substrate is measured (Fig. 13).

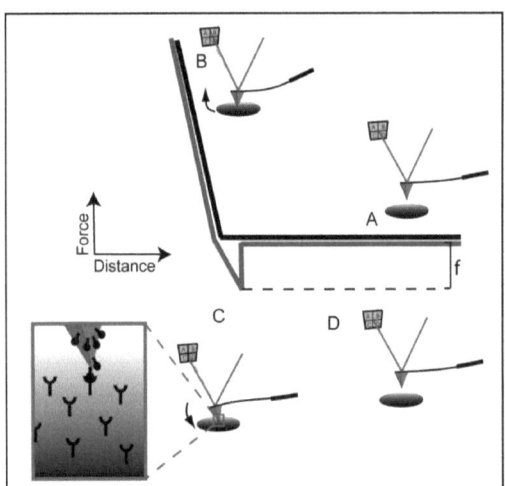

Fig. 13. AFM force spectroscopy. (A) The cantilever tip is brought in contact with a substrate. (B) During contact the tip interacts with the substrate. (C) When the cantilever is withdrawn, the cantilever bends down due to established interactions until (D) the tip is released. By using a cantilever functionalized with a certain receptor and a surface coated with the respective ligand, ligand-receptor inter-actions can be analysed.

First AFM force spectroscopy experiments were conducted with the classical ligand-receptor system Streptavidin-Biotin[174-176], further forces between complementary DNA strands[177], Lectin-lactose[178], antibody-antigen interactions[179-181] interactions. Also cell adhesion receptor interactions have been studied by AFM force spectroscopy, for instance selectin-sialyl Lewis X[182-184], P-selectin-PSGL-1[185-187], Concanavalin A-mannose[188], Cadherin-Cadherin interactions[189, 190] and Integrin-ligand interactions[191]. In the latter examples purified adhesion receptors were analysed. However, as discussed above, a far better approach consists in studying adhesion receptors-ligands interactions in the context of a living cell. This can be easily done with a modified AFM setup. Firstly, AFM needs to be combined with optical microscopy, to ensure well-controlled measurements. Moreover, a temperature-controlled chamber has to be implemented, filled with a sufficiently high volume of cell culture medium to create an appropriate environment for the cells. For certain cell adhesion measurements (especially cell-cell adhesion) an enhanced Z-range of up to 100 μm is further needed. These modifications allow the conduction of AFM-SCFS experiments as detailed below.

2.3.2 AFM- SCFS –experimental setup

In AFM-SCFS experiments adhesion between a living cell and an adhesive substrate, being either another cell or a surface coated with a protein of interest (Fig. 14 A-C) is studied. In cell-surface interaction experiments, the cell is typically attached to an AFM cantilever and interactions with a protein-coated surface are quantified[192-194](Fig. 14 A). Alternatively, the tip can be modified with a protein of interest and probed on a single cell (Fig. 14 A)[195]. The majority of cell-surface interaction studies have applied the aforementioned setup (Fig. 14 A) since it has the advantage of being more versatile with respect to ligand immobilization techniques. Therefore the setup shown in Fig. 14 A will be detailed below.

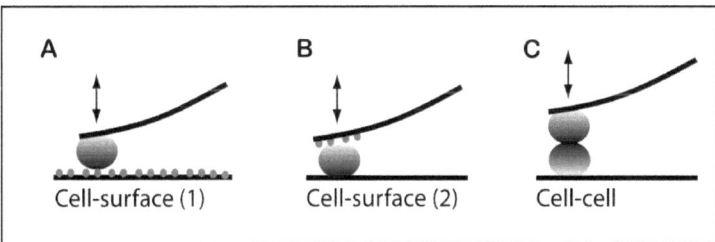

Fig. 14. Possible SCFS experimental setups to measure cell interactions with adhesive substrates. (A) A single cell is immobilized on the AFM cantilever and probed on an adhesive substrate. (B) The cell is sitting on a surface and the ligand-coated cantilever is probed onto it. (C) To quantify cell-cell adhesion, one cell is attached to the cantilever and probed on a cell sitting on the surface.

Cell immobilization on the cantilever

To enable attachment of a living cell to the cantilever, the cantilever surface has to be modified accordingly. Commonly, for eukaryotic cells Concanavalin A (Con A) is used, a lectin that binds carbohydrate groups on the cell surface[196]. There have also been reported alternative methods; for instance wheat germ agglutinin (WGA) has been used to immobilize different cell types[197]. In other studies cells were biotinylated and then bound to a streptavidin-modified cantilever[198] or cells were grown on the cantilever[192]. Furthermore cells can be immobilized via ECM proteins (FN, collagen), RGD peptides or antibodies to AFM cantilevers.

To attach a cell onto the functionalized cantilever, suspended cells are added into the fluid chamber. After the cells settled down, the cantilever is gently brought in contact with a single cell and withdrawn to capture the cell (Fig. 15).

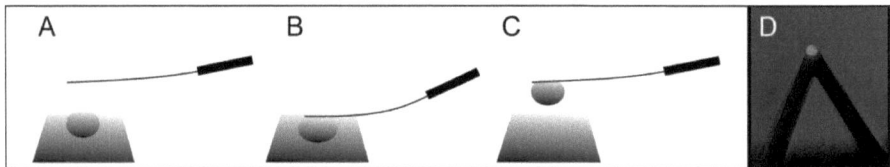

Fig 15. Attaching a living cell onto the AFM cantilever. (A) The functionalized cantilever is positioned above a cell sitting on the substrate and gently pressed onto it (B). During contact, adhesive interactions are established between cell and cantilever coating. (C) Thereafter, the cell-cantilever couple is separated from the surface. During the next minutes firm attachment is established. (D) Green-fluorescent fibroblast (vinculin-GFP) immobilized on a AFM cantilever. The picture is an overlay of images recorded by phase contrast and epi-fluorescence microscopy. The scale bar corresponds to a length of 50 μm.

Recording a F-D curve

Once the cell is stably bound to the cantilever, the cantilever is approached to the surface until a certain force set-point is reached. Thereby the cell is brought into contact with the adhesive substrate. After a given contact period the cantilever is retracted and the cell is pulled away from the substrate (Fig. 16). If adhesive interactions were established between cell and substrate the cantilever bends downwards during retraction until the cell is fully separated from the surface. During the described procedure- the so-called F-D cycle- the force acting on the cantilever is monitored (Fig. 16) (appendix, supplementary videos 1 and 2). Parameters significantly influencing the experiment, such as contact force and period, contact condition (contant height, constant force) and pulling speed can be precisely controlled.

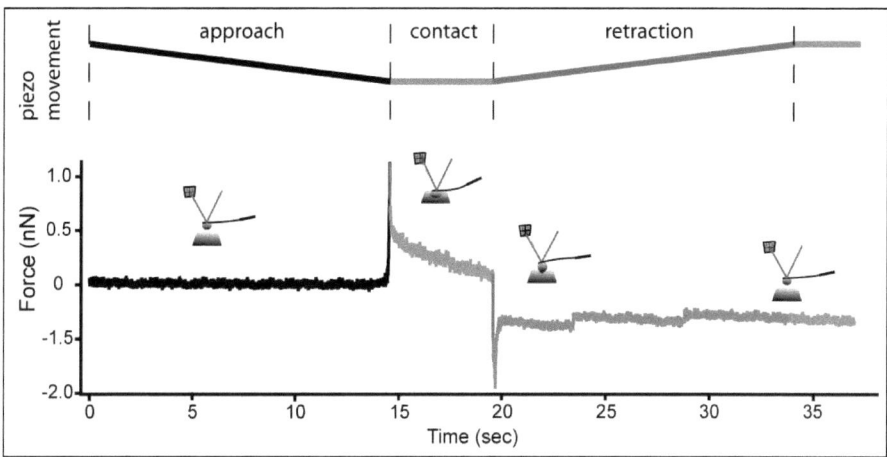

Fig. 16. Monitoring the force signal during a F-D cycle. Piezo movement (above) and force signal (below) versus time during a F-D cycle. The piezo extends until a certain force set-point is reached (black). In constant height-mode the piezo position is maintained constant during contact (grey). Due to its viscous properties, the cell relaxes and the force on the cantilever decays during the first seconds of contact. In the retraction curve (grey) the cantilever bends down due to the established adhesion between cell and substrate. The baseline force level is reached when all linkages between cell and substrate are broken.

Plotting the detected force versus the distance separating cell and substrate generates a F-D curve (Fig. 17).

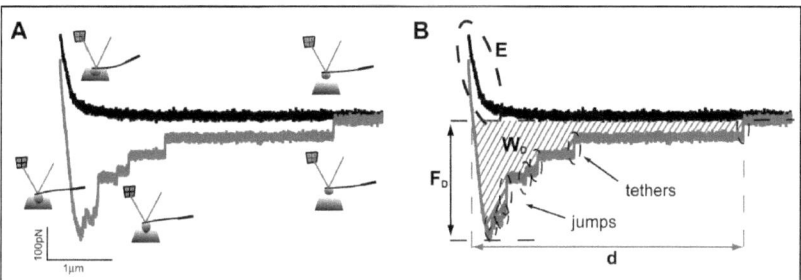

Fig. 17. F-D curve and extracted information. (A) Schematic representation of the F-D cycle. (B) Information that can be obtained from the F-D curve. F_D maximal force required to detach the cell, E elastic properties, W_D adhesion work, d distance required for separation, discrete force steps (jumps and tethers).

2.3.3 Interpretation of SCFS F-D curves

F-D retraction curves (Fig. 17, grey) usually show a characteristic and complex force pattern. How can this force pattern be interpreted? Fig. 18 illustrates different phases of the cell detachment process and correlates these to an example F-D curve.

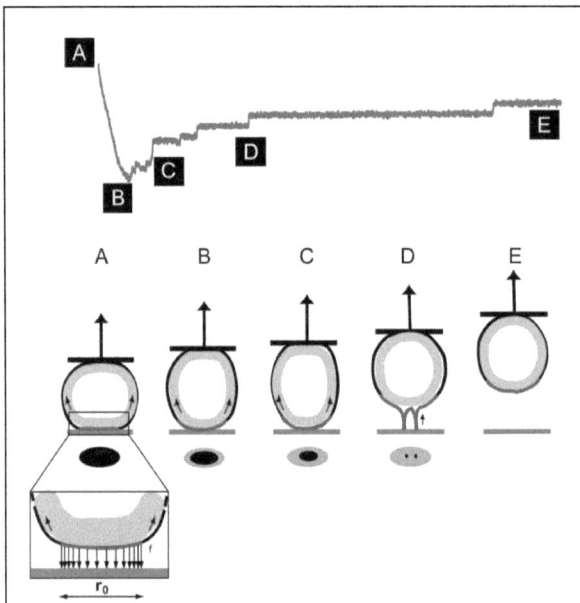

Fig. 18. Schematic representation of the cell detachment process. The cell detachment process is separated into different phases. (A) The cell is in contact with the substrate. In the contact zone adhesive interactions occur. (B,C) During cell detachment established bonds are ruptured and the formed contact zone shrinks. When the cell body is separated from the surface, solely membrane nanotubes link cell and substrate until the cell is fully detached from the substrate (D).

F-D curves do not only contain useful information about the established interactions between cell and substrate, but also give information about the mechanical properties of the cell, for instance. Subsequently parameters that can be extracted from F-D curves are discussed. Thereby *overall adhesion* is defined as the sum of adhesive interactions occurring between cell and substrate.

Detachment force F_D:

The detachment force is defined as the maximal force that is required to separate the cell from the substrate. How can the detachment force be interpreted? By light microscopy it can be observed that the apparent[3] contact zone established between cell and substrate is approximated by a circular area (see Fig. 39 and 46). During the initial detachment phase, bonds established in the outer zone of the contact zone are stressed first. Depending on the established adhesion, the cell is stretched until a maximal force is reached. Upon bond failure, the contact zone shrinks. Assuming a homogenous distribution of receptors over the contact zone, more bonds per radial section will have formed at the periphery of the contact zone than in the inner region (Fig. 18A). Consequently a maximal force is detected initially before the bonds at the periphery begin to fail. After the maximal force is overcome, the force decreases quickly because the applied force load is shared by fewer receptors in the inner contact zone and the probability that these bonds resist rupture decreases[199]. Since the total number of receptors and their binding strengths contribute to F_D, it is most commonly used to quantify overall cell adhesion.

Jumps (j)/tethers (t).

F-D curves usually display small discrete force steps. These can be distinguished into *jumps* (j) and *tethers* (t) (Fig. 17). Whereas j events are typically preceded by a non-linear force loading, a force plateau is detected prior to t (Fig. 19). Both, force gradient prior rupture (>0: j, ~0: t) and the distance at which the force jumps occur (dependent on the context) can be used to distinguish j and t. Distinguishing j and t events is necessary, since they are contributed to different detachment scenarios (Fig. 19):

j events. j events can be interpreted as the rupture of single or a few ligand-receptor bonds. Unbound receptors are supposed to be linked to the cytoskeleton (Fig. 19)[200]. For instance, integrins often localize to specialized complexes involving assemblies of cytoskeletal linker and signalling proteins. Stretching of these membrane-cytoskeleton linkers leads to a non-linear force increase prior bond rupture[200]. The magnitude of the force step reflects the stochastic survival of this ligand-receptor bond under an increasing force load[199, 200]. The ensemble of many j events can provide information on the binding strength (discussed in 3.3).

t events are often found in F-D curves at pulling distances of several μm. The force plateau preceding the force step (t) appears when membrane nanotubes (often also called membrane tethers) are pulled out of the cell membrane (Fig. 19, 20). Membrane nanotubes may form in two different processes: either one or few receptors that are not linked to the cytoskeleton bound their ligands and

---

[3] *Unequal to the molecular contact area in which cell and substrate are in sufficiently close proximity to enable receptor-ligand interactions.*

are pulled –together with a membrane nanotube- from the cell. Alternatively receptor/receptors are pulled from the cell after the molecular link coupling them to the cytoskeleton has been disrupted[200] (Fig. 19).

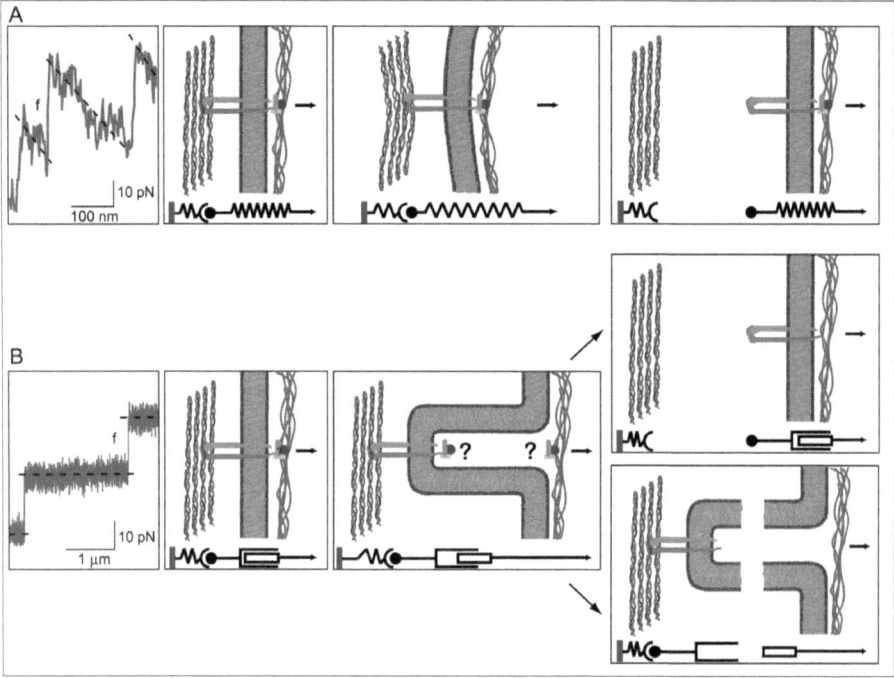

Fig. 19. Sketch illustrating the different events causing j and t events. A. "j"-like events (left). A receptor anchored to the cytoskeleton binds to a ligand in the ECM (here collagen). Upon pulling on the cell during cantilever retraction, the receptor-membrane-cytoskeleton linker is stretched and the force on the cantilever increases. Upon bond rupture the force on the cantilever rapidly decreases. B. "t"-like event (left). A receptor that is not anchored to the cytoskeleton (alternatively anchorage is disrupted during pulling) is extracted with a membrane nanotube from the cell body. The force on the cantilever remains constant during tether extraction. When the receptor-ligand bond is released, the force on the cantilever decreases staircase-like (upper sketch). Alternatively, the nanotube might fail (sketch below) or the receptor might be pulled out of the membrane.

In case of a receptor sustaining the nanotube, the nanotube may be released upon receptor-ligand dissociation. In such case the length of the nanotube can be used to calculate the lifetime of the receptor-ligand interaction at a certain force[201]. Thereby the bond lifetime- approximately equal to the tether survival time- is calculated by dividing nanotube length through the pulling speed. During tether formation the plasma membrane is deformed, thus the magnitude of t events reflects the mechanical characteristics of the probed membrane. These include surface tension and binding rigidity[202-205]. The biophysical aspects of membrane nanotube pulling have been characterized in phospholipid vesicles and different cell types[204-206]. Membrane nanotubes can also be found *in vivo*. They are suggested to facilitate the transfer of vesicles and small molecules from one cell to another[207]. Thus, there exists certain interest in studying the biophysical behaviour of membrane nanotubes. Different to some tethers formed *in vivo* that contain actin filaments or microtubuli, artificially created tethers usually exhibit no cytoskeleton compounds.

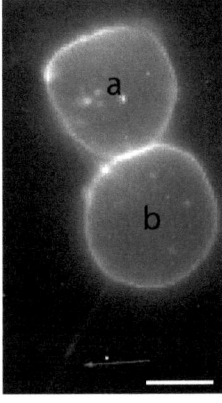

Fig. 20. Image of a membrane nanotube pulled from an embryonic zebrafish cell. A ConA-coated cantilever was brought into contact with a cell having fluorescent labelled membrane (cell b). Upon separating the cantilever from the cell, a membrane nanotube (arrow) is formed between cantilever and cell (the cantilever cannot be seen here). Membrane nanotubes usually exhibit a radius of approximately 100 nm and can have a length of several tens of μm. Scale bar corresponds to 10 μm.

Separation distance d,

The separation distance is the distance at which all linkages between cell and substrate between cell and substrate have been ruptured. This length is highly influenced by membrane nanotubes.

Elastic properties of the cell E.

The slope of the force increase in the approach curve during contact formation (Fig. 17) is influenced by the elastic properties of the cell. The elastic properties of most cell types are substantially influenced by organized actin filaments of the cell cortex. Fig. 21 shows F-D curves (approach) recorded in the presence of increasing concentrations of *Cytochalasin D* that destabilizes actin filaments. The cell becomes softer with increasing *Cytochalasin D* concentration in the medium, as indicated by the increasing cell deformation (0.25 μm - 0.5 μm – 1 μm). Moreover, during contact with the substrate, a decay of the force can be observed in constant height mode (Fig. 16). This decay is explained by the viscous properties of the cell. Approach F-D curves can be used in appropriate experimental setups to analyse the viscous properties of cells[208, 209] (see below).

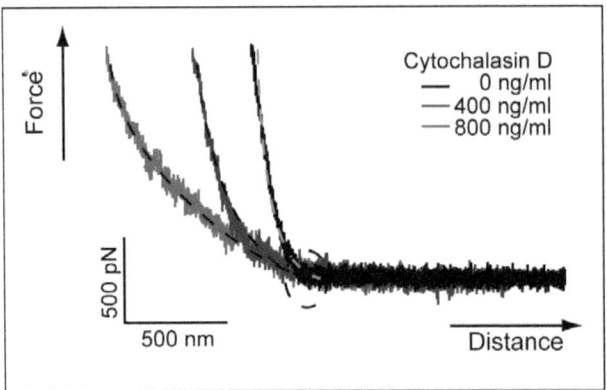

Fig. 21. F-D curves recorded in presence of cytochalasin D. Only approach curves are shown. F-D curves were recorded after adding increasing concentrations of Cytochalasin D to the medium. The circle illustrates the region in which contact is established between cell and surface. Trend fits describe the non-linear force increase during indentation.

Detachment work W_D

The detachment work corresponds to the work that has to be done to detach the cell from the substrate. W_D is found by measuring the area enclosed by retraction curve and baseline (hatching in Fig. 17). W_D relates to the overall adhesion established between cell and substrate, but is also substantially influenced by the elastic properties of the cell (see covariance data, Appendix A1). Due to their length, often ten´s of μm, membrane nanotubes significantly contribute to the W_D.

2.3.4 SCFS –state of the art

With the described setups various experiments can be designed. Advantageously AFM based SCFS allows forces within a large range from a few pN up to several hundreds of nN to be measured. Thus, interactions mediated by single CAMs [153, 193, 198, 210, 211] up to adhesive interactions established by larger adhesive complexes can be detected[197, 212]. This versatility makes it possible to address a broad range of biological questions.

Pioneering SCFS experiments with living mammalian cells were carried out by *H. Gaub´s* group[192], quantifying cell-cell adhesion between trophoblasts and uterine epithelial cells, to model interactions occurring during embryo implantation. In another early study, Lehenkari et al. analysed adhesion between osteoblasts/osteoclasts and different RGD-containing ligands[195].

Numerous SCFS experiments were performed by *V. Moy*'s group, for instance characterizing $\alpha_5\beta_1$-integrin-FN interactions at the single-molecule level[193]. Furthermore, overall endothelial cell-leukocyte adhesion[213, 214] was investigated and the contribution of integrin- and selectin-mediated interactions in overall adhesion was demonstrated by using respective blocking antibodies. These experiments could provide insights into the mechanisms underlying adhesive interactions between leukocytes and endothelium that are crucial to initiate the process of transmigration during inflammatory response. Adhesive interactions between endothelial cells and leukocytes interactions involve LFA (lymphocyte function-associated molecule)-1-ICAM (intercellular adhesion molecule)[196, 215] and $\alpha_4\beta_1$-integrin-VCAM (vascular cell adhesion molecule)[216] interactions; these were further explored at the single molecule level. Thereby bond specific parameters, such as bond dissociation rates could be determined. Recently homophilic JAM-A (junctional adhesion molecule) interactions were characterized and a role of LFA-1 in their regulation was shown[217].

For several years, the group of *D. Müller* has established AFM-SCFS as a tool to quantify cell adhesion. An important improvement of experimental setup was the implementation of an enhanced piezo range that enables to perform sensitive cell-cell adhesion measurements[218]. This optimized setup further allowed probing adhesion over a broad range of detachment forces, starting with single-molecule interactions up to high forces exerted by more complex adhesive sites[219]. Moreover, by using a system combining SCFS with advanced optical microscopy techniques, a better control of the experiment was provided. In first projects adhesion of gastrulating zebrafish cells to FN coated surfaces has been quantified. Thereby the role of *Wnt11* in modulating integrin-mediated adhesion was investigated. *Wnt11* -deficient mutants showed decreased adhesion to FN, which was attributed to decreased integrin binding[194]. In a similar setup the impact of *Wnt11* on intercellular adhesion of gastrulating zebrafish cells was studied. Thereby *Wnt11* was found to modulate E-cadherin mediated adhesion via a Rab5-dependent mechanism[220]. Furthermore adhesion between gastrulating zebrafish cells raised from different germ layers[221] was compared. Thereby mechanisms underlying cellular movements during gastrulation could be investigated and the contribution of differential cell adhesion and cell cortex tension in germ layer organization could be deciphered. In two further studies the role of galectin-3 in adhesion to laminin-111 and collagen type I has been quantified. Thereby galectin-3 was demonstrated to inhibit cooperative $\alpha_2\beta_1$-integrin binding on collagen type I[222, 223]. In another study the role of the integrin activator TPA in strengthening integrin-cytoskeleton interactions and increasing $\alpha_2\beta_1$-integrin avidity could be demonstrated[224]. For all these projects principally the same SCFS setup could be used, only with slight modifications (appendix B1.1, table S2).

In *D. Wirtz's* group homophilic and heterophilic N-, E- and VE-cadherin interactions were further characterized at the single-molecule level[198, 225]. In further experiments the role of alpha-catenin in strengthening cadherin-mediated interactions was examined[226]. Other groups analysed interactions between isolated P-selectins and their ligands on the surface of polymorphonuclear leukocytes and colon carcinoma cells at the single-molecule level[227]. Furthermore, adhesive interactions of cell surface integrins with nanopatterned RGD-peptide coated substrates were quantified. Thereby insights into the optimal spatial organization of RGD ligands for integrin binding could be gained[212].

Alternative SCFS setups

SCFS can also be used in alternative experimental setups to address different biological problems. SCFS can probe mechanical properties of cells, for instance (Fig. 22 A): Thereby the AFM tip or a bead attached to the cantilever is used to indent a cell in a controlled manner. The indentation resulting from a given force or the force required to reach a certain indentation depth can be measured and compared among different cell types/conditions. In a more quantitative approach the relationship between force and indentation is used to determine the cell's young modulus according to Hertz model[208, 209, 228-230] or alternatively the *liquid droplet* model[228]. By combining imaging and elasticity measurements, local elastic properties of a cell can be analysed and an elastic map of the cell can be created. Such experiments were used to assess local or temporal changes in elastic properties of a cell upon certain treatment, for instance anti-cytoskeletal drugs, chemotherapeutic drugs or radiation[231-234].

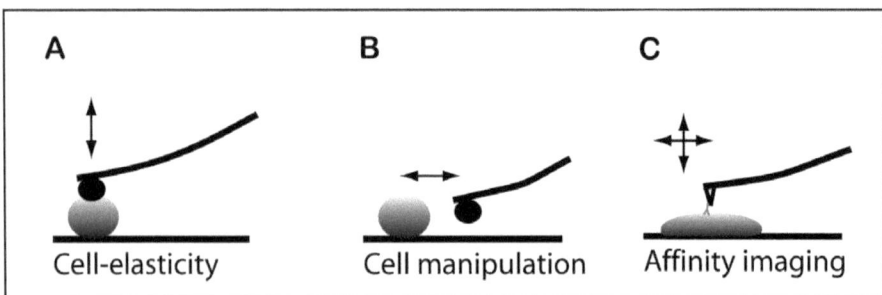

Fig. 22. Alternative SCFS experimental setups. (A) Studying mechanical properties of cells. (B) Manipulating cells at high temporal and spatial precision. (C) Creating an interaction map of living cells using a receptor/ligand- functionalized AFM tip.

The described experiments can give valuable insights into altered elastic properties of cells, which is relevant to several diseases, for instance in osteoarthritis and in cancerous diseases, e.g. leukemia [228]. Several independent studies have for instance reported that malignant cancer cells are more compliant compared to non-malignant cells[228, 235-237].

Moreover, AFM cantilevers may be used to specifically manipulate cells in multiple ways (Fig. 22). This allows investigating the behaviour of cells when these are subjected to mechanical stress. In a previous work, for instance, osteoblasts were indented by a bead attached to an AFM cantilever; by using an colorimetric assay in parallel, the mechanical stimulus was directly shown to cause an increase of intracellular calcium concentrations [238]. Furthermore forces generated by cells can be quantified, for instance during cell migration[239, 240] or during mitotic cell rounding.

Other applications include affinity imaging (Fig. 22 C), in which an array of F-D curves is recorded on a surface, for instance the surface of a cell. By using a cantilever tip decorated with specific bio-molecules (e.g. an antibody, specific ligand/receptor) an interaction map of this surface can be generated[241-244 181, 244, 245].

2.4 Conclusions

As shown by the given examples, AFM-SCFS is a versatile tool that allows the performance of sensitive cell-ECM and cell-cell adhesion measurements at a high temporal precision. Advantageously a large range of adhesion forces can be detected. The high force resolution further allows the analysis of single-molecule interactions. How can the measured forces be interpreted? To extract biophysically relevant data, e.g. information about bond dissociation rates, a model is needed that allows to relate measured forces to binding rates. In the following chapter, chapter three, the Bell-Evans model that was used in the presented work to interpret force spectroscopy data is introduced.

Chapter 3

The *Bell-Evans* model

3.1 Basic reaction equations & bond kinetics for receptor-ligand interactions

About thirty years ago a seminal model was proposed by George Bell to describe the kinetics of receptor-mediated adhesion occurring between two cells or between a cell and a ligand-coated surface[246]. Expanded by Evan Evans, this model deals with the effect of an externally applied force on the lifetime of biomolecular bonds. Thus, it provides an appropriate framework to interpret data obtained in above described AFM-SCFS experiments at the molecular level.

In AFM-SCFS experiments such as conducted in the presented work, a single cell bound to the AFM cantilever is brought into contact with a surface coated with ligand molecules (Fig. 23). Upon reaching contact with the surface, the cell deforms and contact is established over a certain contact zone. Then the cell is held in contact for a certain period. Within the contact zone interactions between the cell and the surface can occur, these include unspecific and specific interactions. Thereby unspecific interactions are defined as interactions that occur between all cells, for instance ionic interactions[*], steric repulsion and van der Waals forces[246]. In contrast, specific interactions are by definition mediated by receptors (CAMs) that bind to their ligands on the surface. The binding reaction in the contact zone can be written as:

$$R + L \underset{k_{off}}{\overset{k_{on}}{\rightleftharpoons}} RL \qquad\qquad Eq.\ 1$$

where R are the free receptors, L the free ligands, k_{on} and k_{off} are the rate constants for formation and dissociation of the bound state RL respectively [246].

---

[*] *The lipid bilayer of cells has net negative charge, thus repelling forces act among cells. In physiological medium this force is mostly reduced by surrounding counter- ions, thereby the Debye length has been estimated to be in the order of 1nm in physiological media. Furthermore, ionic interactions occur among charged sidegroups of cell- surface proteins*[246].

These receptor-ligand interactions can occur between binding partners that are in close proximity with each other. Since the CAMs are inserted in the cell membrane, the distance between cell membrane and the plane in which the ligand is immobilized has to be small enough, such that a ligand and a receptor can bind. The proximity in vertical direction is established by approaching the cell by the cantilever movement and further by cellular fitting to the surface during contact. Both processes also reduce steric repulsion mediated by large extracellular molecules of the glycocalyx[247]. For simplicity it is assumed that the receptors are initially uniformly distributed throughout the contact area[246].

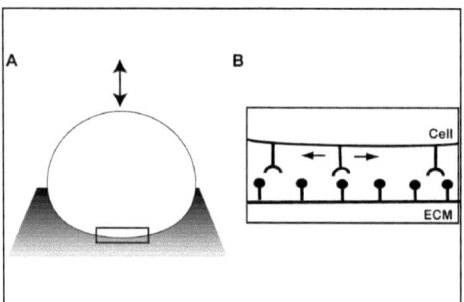

Fig. 23. Sketch illustrating the contact zone established between a cell and a ligand-coated surface during AFM-SCFS. The initially applied force establishes contact and brings receptors and ligands together. Bonds can form if receptors and ligands are in close proximity to each other. Receptors can diffuse in the membrane (see arrows).

Once contact is established, receptors and ligands encounter each other because the receptors are motile and diffuse in the plane of the cell membrane. Diffusion constants of CAMs have been measured by various techniques, for instance by fluorescence recovery after photo-bleaching (FRAP), fluorescence correlation spectroscopy (FCS) or by single-particle tracking (SPT)[248-250]. Depending on the analysed system, diffusion constants between $10^{-11} - 10^{-9} m^2/\text{sec}$ have been found[250, 251]; they are influenced by the motility of the proteins and interactions with other proteins or linkage to the cytoskeleton. For instance, for integrin $\alpha_4\beta_1$ a diffusion constant of $D = 3 \cdot 10^{-10} m^2/\text{sec}$ has been measured when not cytoskeleton-associated, and $D = 5 \cdot 10^{-11} m^2/\text{sec}$ when linked to the cytoskeleton[249]. Differently to CAMs, ligand molecules immobilized on the surface are not free to move. Since the diffusion constants are approximately four orders of magnitude smaller for membrane-inserted proteins than for proteins free in solution, the bond formation rate k_{on} will be near the diffusion limit[246]. The small diffusion rates further imply that dissociated receptor and ligand can recombine before diffusing apart[246]. Thus, the following kinetic equations can be formulated to describe the time-dependent change of the number of ligand-receptor interactions per unit area N_{RL} (unit: $1/\mu m^2$):

$$\frac{d[I_{RL}]}{dt} = k_{on} \cdot N_{Rf} \cdot N_{Lf} - k_{off} \cdot N_{RL} \qquad Eq.\ 2$$

with N_{Rf} and N_{Lf} being the number of free receptors and ligands, respectively, per unit area. With N_R and N_L being the total number of receptor/ligands per unit area, it can be written:

$$N_R = N_{RL} + N_{Rf} \qquad Eq.\ 3$$

and

$$N_L = N_{RL} + N_{Lf} \qquad Eq.\ 4$$

Thus, *Eq. 2* can be rewritten as follows[246]:

$$\frac{d[I_{RL}]}{dt} = k_{on} \cdot (N_R - N_{RL}) \cdot (N_L - N_{RL}) - k_{off} \cdot N_{RL} \qquad Eq.\ 5$$

Formulating *Eq. 5* it was assumed by G. Bell that the overall receptor density within the contact area remains constant[†]. In AFM-SCFS experiments the cantilever is withdrawn after a certain time period. During separation the force is recorded for each distance separating cell and surface. If the cell remains attached to the substrate, a linearly increasing force is exerted onto the ligand-receptor bonds $F = k \cdot x$. How does this force affect the unbinding rate of receptor-ligand bonds?

3.2 Dissociation kinetics *near* and *far* from equilibrium

Interactions between CAMs and their ligands are governed by weak non-covalent interactions such as van der Waals, electrostatic, hydrophobic/hydrophilic and hydrogen bond interactions. Dissociation kinetics is determined by the energy landscape of barriers, being the free energy profile along a preferential pathway (there are more possible pathways, which might be traversed, but these might be energetically less favorable)[253]. This energy landscape is described by a minimum (bound state) that is separated by one or several energy barriers from the unbound state (Fig. 24). These potential barriers influence the kinetics of bond dissociation in a way that the bond lifetime τ increases exponentially with the barrier height:

$$\tau = \tau_0 \cdot \exp\left(\frac{E_b}{k_B \cdot T}\right) \qquad Eq.\ 6$$

[†] A possible change of receptor density might occur due to inward-diffusion of membrane bound receptors outside the contact area[252] or by transport of new receptors from the cytoplasm to the cell surface.

where τ is the bond lifetime ($\tau = 1/k_{off}$), τ_0 is the reciprocal of a natural frequency of oscillations, E_b the height of the energy barrier, k_B the Boltzmann constant and T the absolute temperature[254].

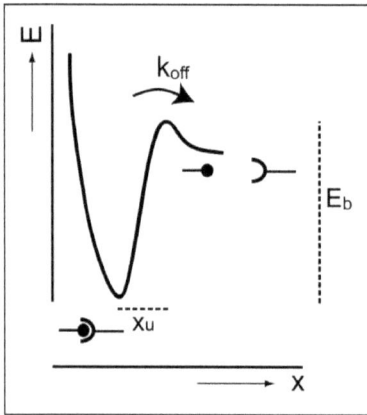

Fig. 24. Schematic representation of receptor-ligand bond dissociation. *A potential barrier (height E_b, width x_u) has to be overcome to dissociate a ligand-receptor bond. . For simplicity, a model with a single-barrier is shown. The energy landscapes of biomolecular bonds is expected to be much more complex, since many interaction sites are involved[255]. Figure modified from[255].*

How does an externally applied force affect bond lifetime? Adhesive bonds often function under mechanical stress[246, 256, 257]. Good example are leukocytes that experience shear stress while they are attached to blood vessels. Furthermore, pulling forces are exerted on adhesion sites anchoring fibroblasts during wound contractions and muscle cells during muscle contraction[256]. Also cells within static tissues can transiently be subjected to high mechanical stress. Such forces determine substantially bond lifetimes[246]. In his model describing cell-cell adhesion, G. Bell proposed the following relationship between bond lifetime τ and force[246]:

$$\tau(f) = \tau(0) \cdot \exp\left(\frac{x_u \cdot f}{k_B \cdot T}\right) \qquad Eq.\ 7$$

or analogously, substituting τ by $1/k_{off}$ and $\tau(0)$ by $1/k_{off}(0)$

$$k_{off}(f) = k_{off}(0) \cdot \exp\left(\frac{x_u \cdot f}{k_B \cdot T}\right) \qquad Eq.\ 8$$

where τ(0) is the lifetime under zero force; $k_{off}(0)$ the dissociation rate under zero force, k_B the Boltzman constant and T the absolute temperature, f is the applied force and x_u is the width of the energy barrier.

According to *Eq. 7* a force f applied to the ligand-receptor interaction lowers the energy barrier by $x_u \cdot f$ (Fig. 25), and shortens the bond lifetime by the factor of $\exp\left(\frac{x_u \cdot f}{k_B \cdot T}\right)$. Since the lifetime is the reciprocal of the bond dissociation rate k_{off}, it follows that k_{off} increases exponentially with the applied force (*Eq. 8*). Thus, a single bond resists to a force for a time period that is smaller than the time needed for its spontaneous dissociation under thermal activation $\tau(0)$[256]. The relationship in *Eq. 7* was taken over by Bell from solid-materials theory. A similar relationship was found experimentally in fraction tests with various solid materials, ranging from metals to polymers[258]. By applying this relationship between force and lifetime to the case of the interaction of two adhering cells, G. Bell postulated that the dissociation of biological bonds had similarity to the rupture of solid materials. The predictions of G. Bell were later experimentally and theoretically confirmed by E. Evans and corrected to describe the force-induced rupture of ligand-receptor interactions[254].

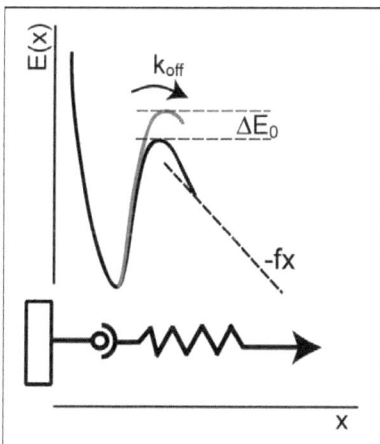

Fig. 25. Influence of an external force on a single energy barrier separating the bond and the unbound state. *Along the reaction coordinate x, an external force adds a mechanical potential –fx that tilts the energy landscape and lowers the barrier to dissociation (figure modified from* [256]*).*

As illustrated in Fig. 25, a force that is applied via a structural element with an effective spring constant k_{eff} on a molecular bond creates a spring-like potential that alters the chemical energy of interaction.

Substituting the dissociation rate from Eq. 8 into *Eq. 5*, the kinetics of bond formation under force can be newly formulated as[246]:

$$\frac{d[I_{RL}]}{dt} = k_+ \cdot (I_R - N_{RL})(N_L - N_{RL}) - N_{RL} \cdot k_- \cdot \exp\left(\frac{u \cdot F/C}{k_B \cdot T}\right) \qquad Eq.\ 9$$

where F/C is the force applied to a single bond. Above a certain critical force rebinding events (=first term) can be neglected. Thus, dissociation under sufficiently high force represents kinetics far from equilibrium[246, 254].

According to *Eq. 9* an ideal situation is considered, in which the applied pulling force is equally distributed among all receptor-ligand interactions. This formula could theoretically allow calculating the kinetics of failure of multiple bonds as occurring in our SCFS experiments. In most physiological situations multiple bonds are formed between cells and their environment. These include either multiple interactions by the same type of receptor, or interactions mediated by different receptors. However, the unbinding of multiple receptors is non-trivial: even given the (improbable) situation that only multiple interactions of a single type of receptor, for instance a certain integrin heterodimer, are analysed, the situation could become very complex: some heterodimers might not be linked to the cytoskeleton, others are bound to cytoskeletal proteins, others might be located in dense adhesion receptor clusters. Consequently local elasticity varies and therefore the individual bonds do not evenly share the applied force. The mentioned aspects might have as a consequence that a broadened distribution of rupture forces is detected, furthermore multiple distributions could appear.

This implies that single molecule interactions have to be analysed. Therefore the number of occurring interactions has to be reduced. In AFM-SCFS experiments, this can be achieved by reducing both, contact time and contact force to minimize the established contact area. Under such conditions, F-D curves with only one single-rupture event can be obtained (Fig. 26).

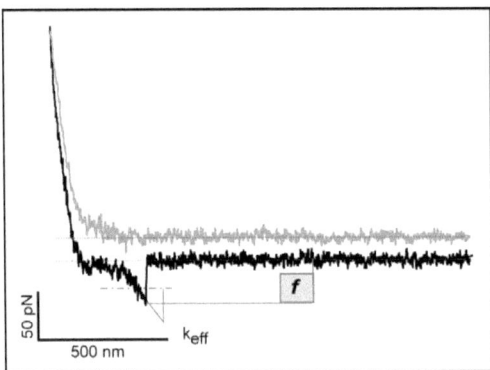

Fig. 26. Representative F-D curve recorded in AFM-SCFS experiments. To reduce integrin binding to single-molecules, short contact time and low contact force was chosen. From the F-D curve the rupture force (f) and the spring constant of the linker k_{eff} can be extracted.

To test that single-rupture events are due to specific receptor-ligand unbinding, respective control experiments have to be conducted, for instance by using specific blocking antibodies or by comparing different mutants (see chapter four). From a set of F-D curves a distribution of rupture forces is obtained. This distribution contains information on the binding strength of the probed ligand-receptor interaction as discussed below.

3.3 Binding strength

The binding strength f^* is defined as the most probable[254] or mean[259] force that produces bond failure in repeated detachment tests. For a single receptor- ligand bond subjected to a force f(t), the probability P(t) that it survives in its bound state is given by[254]:

$$\frac{dP(t)}{dt} = -k(f(t)) \cdot P(t) \qquad Eq.\ 10$$

Since bond rupture is a stochastic event, f^* is dependent on the rate by which force is applied to the receptor-ligand interaction (the loading rate r_{eff}). It was theoretically derived and experimentally shown by Evans and Ritchie[254] that f^* increases linearly with the logarithm of the loading rate (Fig. 27):

$$f^* = \frac{k_B T}{x_u} \cdot \ln(\frac{r_{eff} \cdot x_u}{k_B T \cdot k_{off}}) = \frac{k_B T}{x_u} \cdot \ln(\frac{x_u}{k_B T \cdot k_{off}}) + \frac{k_B T}{x_u} \cdot \ln(r_{eff}) \qquad Eq.\ 11$$

Eq. 11 predicts f^* to increase linearly with the logarithm of the applied force loading rate r_{eff}. In force spectroscopy experiments the force loading rates are varied by applying different pulling speeds (see B1.3). f^* is determined for each dataset recorded at a certain pulling speed (corresponding to an average r_{eff}) and plotted versus $\ln(r_{eff})$ (Fig. 27 A, right and 4.3.2). By fitting data with Eq.11, x_u and k_{off} can be calculated.

In previous experiments several integrin-ligand interactions have been characterized in AFM-SCFS experiments, for instance $\alpha_5\beta_1$-FN[261], $\alpha_4\beta_1$- VCAM-1[262] and $\alpha_L\beta_2$- ICAM-1 and $\alpha_L\beta_2$- ICAM-2 interactions[263]. In addition different cadherin interactions were analysed by SCFS, e.g. VE-cadherin[264] and N-cadherin interactions[265]. In all these mentioned cases the Bell-Evans model could be used to interpret the relationship between binding strength and force load and to determine k_{off} and x_u.

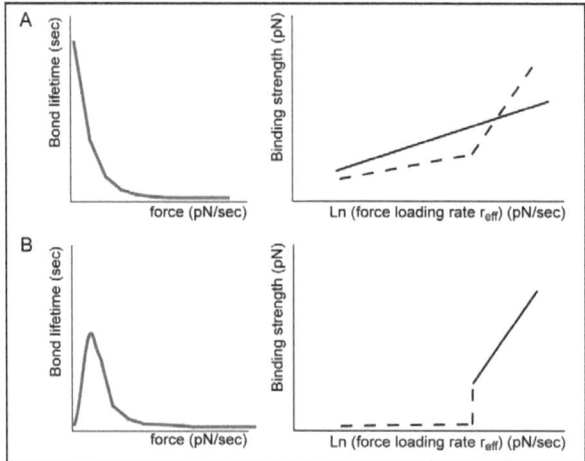

Fig. 27. Schematic presentation showing the effects of force on average bond lifetime and binding strength of slip and catch bonds. (A) Slip bond. The exponential decrease of the bond lifetime- as predicted by the Bell model- with increasing force is shown (left). (right) Linear increase of the binding strength with the logarithm of the loading rate. The solid line shows the behaviour for a single-potential barrier overcome during unbinding, as illustrated in Fig. 24. The dashed line corresponds to the force increase for two potential barriers. (B) Catch bonds. In the low loading rate regime, the average lifetime increases with applied force until a maximum is reached. Thereafter the average lifetime decays (left). (right) Effect of increasing force loading rate on the binding strength. The binding strength of a catch bond is switched at a fast loading rate. Modified from [256] and [260].

However, there are some examples of ligand-receptor bonds in which the Bell model cannot be used for interpretation. In fact, it has been found that the effect of force on bond lifetimes can be very different. Some ligand-receptor systems show quick unbinding in the low-loading rate regime and strengthen their binding -within a certain range- with augmenting force loading. Bonds showing such behaviour are termed "catch bonds"[257] (Fig. 27 B). Catch bonds have been reported for certain types of adhesion receptors, for instance p-selectin PSGL[257, 266-268, 269] and the most common *E. coli* adhesin, FimH[270, 271].

In the case of p-selectin, the catch bond behaviour has been proposed to be adapted to the physiological environment of leukocytes expressing those molecules. In the blood stream, leukocytes are exposed to varying shear rates. Leukocytes show a shear-threshold behaviour; they move freely below a certain shear stress, at increased shear rate they start rolling on blood vessels due to transient binding events[260]. Thus, catch bonds can modulate cell adhesion under varying mechanical stress[257].

3.4 Conclusions

Experiments conducted far from equilibrium are of particular interest for the study of ligand-receptor bonds that experience forces *in vivo*, as basically all CAM-ligand bonds do. Since the behaviour of adhesion receptors under external forces can be such diverse as described, the relationship between bond lifetime and external force has to be experimentally tested for each ligand-receptor system. As seen in the previous section, AFM-SCFS represents an optimal tool for that purpose, since adhesion bonds are probed in their native environment over a large range of loading ranges. Thereby energy barriers that are essential to the dynamic functions of adhesion molecules may be detected[253].

Chapter 4

Quantifying early steps of $\alpha_2\beta_1$-integrin mediated cell adhesion to collagen type I

4.1 Abstract

In this first project early steps of $\alpha_2\beta_1$-integrin mediated cell adhesion to two-dimensional collagen type I matrices (Col) were quantified by AFM-SCFS. CHO cells that had been transfected with α_2-integrin subunits (CHO-A2) were chosen for the analysis. α_2-integrin negative CHO-WT served as controls. For all tested contact times, detachment forces were significantly higher for CHO-A2 cells compared to CHO-WT, suggesting that adhesive interactions between CHO-A2 cells and Col were dominated by $\alpha_2\beta_1$-integrin-collagen bonds. Using CHO-A2 cells, $\alpha_2\beta_1$-collagen interactions were subsequently analysed at the single-molecule level. Dynamic force spectroscopy permitted calculation of bond specific parameters, such as the bond dissociation rate k_{off} (1.3±1.3 sec^{-1}) and the barrier width x_u (2.3±0.3 Å). Next, the time-dependent build-up of CHO-A2 cell adhesion was monitored over 600 sec. Detachment forces increased slowly during the first 60 sec and the single rupture forces were consistent with the unbinding of individual integrin-collagen bonds. When contact between cell and Col was sustained for > 60 sec, a fraction of cells suddenly reinforced adhesion and detachment forces increased up to tenfold. Interestingly, adhesion reinforcement coincided with a rise of the individual rupture events above the values measured for individual $\alpha_2\beta_1$-integrin-collagen bonds. This suggested that integrin clusters had formed which allowed cooperative integrin binding. When adhesion was quantified in the presence of inhibitors of acto-myosin contractility less cells reinforced adhesion. This indicated that acto-myosin contractility was required for the establishment of cooperative integrin binding. In conclusion, the kinetics of $\alpha_2\beta_1$-integrin mediated adhesion was investigated and insights into the underlying binding mechanisms were obtained.

4.2 Introduction

Within the integrin family, collagen binding integrins present a structurally and functionally distinct subgroup comprising four integrin heterodimers, $\alpha_1\beta_1$, $\alpha_2\beta_1$, $\alpha_{10}\beta_1$ and $\alpha_{11}\beta_1$[272, 273]. These collagen receptors share a common β_1-integrin subunit that can associate with the four structurally related α-integrin subunits α_1, α_2, α_{10} or α_{11}[1]. The expression of collagen binding integrins varies among different cell types. $\alpha_1\beta_1$- and $\alpha_2\beta_1$-integrins are widely expressed collagen receptors. For instance, $\alpha_1\beta_1$-integrin is present on smooth muscle cells and in many mesenchymal cell types[276]. $\alpha_2\beta_1$-integrin was first identified in 1987 in the human fibrosarcoma cell line HT-1080[274] and later found to be abundant on epithelial cells, platelets and mesenchymal cells[277, 278]. $\alpha_{10}\beta_1$- and $\alpha_{11}\beta_1$-integrins are more recently discovered integrin heterodimers and compared to $\alpha_1\beta_1$ and $\alpha_2\beta_1$, their expression is more limited. Whereas $\alpha_{10}\beta_1$ is specifically expressed in cartilage[279]. $\alpha_{11}\beta_1$ is expressed in fibroblasts[280] and in many mesenchymal tissues during development[273, 281].

$\alpha_1\beta_1$- and $\alpha_2\beta_1$-integrins transmit different signals into the cell and thereby fulfill different biological functions. Specific signalling pathways were revealed in experiments using α_1- and α_2 knockout mice and by using α_1- and/or α_2-negative cell types that were transfected with the respective integrin subunit[273]. Surprisingly the phenotypes of $\alpha_1\beta_1$ and $\alpha_2\beta_1$ knockout mice are mild, suggesting that their function can be compensated by other collagen binding integrins[282]. $\alpha_1\beta_1$-integrin promotes cell proliferation[283] and is a negative feedback regulator of collagen synthesis *in vitro* and *in vivo*[282, 284]. It has a further functional role in inflammatory cells[273]. In contrast, $\alpha_2\beta_1$-integrin activates collagen synthesis and induces expression of matrix metalloproteinases[285]. In fibroblasts and osteogenic cell lines, $\alpha_2\beta_1$ can mediate cell migration on collagen and reorganization of collagen fibrils. When cells are enclosed in three-dimensional collagen gels, $\alpha_2\beta_1$-dependent gel contraction can be observed. In contrast, the effects of $\alpha_{10}\beta_1$ and $\alpha_{11}\beta_1$ on intracellular signalling pathways are not as well known so far. $\alpha_{11}\beta_1$-integrin has been associated with cell migration[286, 287] and collagen fibril assembly[288]. In some cell types, such as fibroblasts, chondrocytes, osteoblasts, endothelial cells and lymphocytes, $\alpha_1\beta_1$ and $\alpha_2\beta_1$ can be expressed at the same time[289, 290]. Since $\alpha_1\beta_1$ and $\alpha_2\beta_1$ can activate opposing signalling pathways, their binding is inversely regulated, as recently observed in renal cells. There are also cell types, for

---

[1] *Whereas initially $\alpha_3\beta_1$ had been initially also described as a collagen binding integrin*[274]*, it has been later found that it might have a more indirect role as an assisting rather than primary collagen receptor*[275]*.*

instance KHOS-240 osteosarcoma cells, in which all four collagen binding integrins are expressed at the same time[290].

All collagen binding integrins contain I-Domains in their α-subunits which are used for ligand binding (see 1.2.2). I-domain crystal structures of $α_1$- and $α_2$-integrins have been solved; they share many similarities. The binding preferences of different collagen binding integrins were elucidated by cell attachment assays using transfected cells or by binding assays performed with purified integrin I-domains. For instance CHO cells which lack endogenous collagen binding integrins were transfected to express either $α_1$- or $α_2$-integrin and attachment and spreading of these cells was analysed on collagen coated surfaces[289]. It became evident that $α_1β_1$ and $α_2β_1$ have different preferences regarding the collagen subtype. $α_1β_1$-integrin expressing CHO cells could spread on collagen types I, III, IV and V, but not on collagen type II[289]. $α_2β_1$-integrins mediated spreading on collagen types I-VI[289]. Although $α_1β_1$ could mediate cell adhesion and spreading to collagen type I[285, 291], binding affinities were significantly lower than for $α_2β_1$[289, 292]. This is in agreement with several studies showing that $α_1β_1$ is a high affinity receptor for collagen type IV[293, 294], whereas $α_2β_1$-integrin was proposed as main integrin receptor for fibrillar collagen type I[292]. Cell attachment experiments with cells expressing $α_1/α_2$ chimeric heterodimers suggested that the different binding preferences of $α_1$- and $α_2$-integrins towards collagen type I are determined by their integrin I-domains[295]. This was confirmed by binding assays with isolated αI-domains showing that $α_2$I exhibited higher binding affinity towards collagen type I compared to $α_1$I[296]. Binding affinities of $α_{10}$- and $α_{11}$-integrins have also been studied. These studies suggest that $α_{10}β_1$ resembles more $α_1β_1$ regarding its binding preferences and that $α_{11}β_1$ has similar binding behaviour as $α_2β_1$[291, 296]. However, results obtained with isolated I-domains are not always consistent with cell binding assays[289]. This suggests that integrin-binding experiments should be analysed in living cells (see 2.2).

Different types of integrin-binding sites in collagens have been proposed. First, there are specific motifs inside the triple-helical areas that can be recognized by collagen receptors. Second, there are cryptic binding sites which will be discussed in the next chapter. Last, there are also binding sites for integrins in the non-collagenous domains[273]. Studying attachment of different $α_1β_1$- and/or $α_2β_1$-integrin expressing cells to collagen type I fragments indicated the presence of distinct binding sites within collagen type I for these integrins[275]. Knight et al. identified GFOGER within the triple-helical domains of collagen type I as a high-affinity binding motif for $α_1β_1$ and $α_2β_1$-integrins[297]. The crystal structure of a triple-helical peptide comprising the GFOGER motif in complex with the $α_2$I-domain has been resolved and provided insights into the $α_2β_1$-collagen binding mechanisms[298] (See 2.1.1). However, it is not known how many different recognition sites for integrins exist in total in collagen type I[273]. In the case of $α_2β_1$, various low-affinity binding sites have been suggested apart from the high-affinity binding motif GFOGER[273]. Single molecule experiments detecting the binding strength to collagen type I may give insights into different binding sites. So far, only few studies have analysed quantitatively integrin- ligand binding at the single molecule level. There has been only one study addressing $α_2β_1$-integrin-collagen type I interactions. Adhesion of keratinocytes to collagen I was studied in a flow chamber and a bond dissociation rate (k_{off}) of $0.06s^{-1}$ was determined[299].

In adherent cells, integrins are usually arranged into highly-organized structures, such as focal complexes or focal adhesions[300] (see 1.2.3). Whereas the molecular composition of adhesion sites and signalling pathways controlling their macroscopic assembly or disassembly have been analysed in detail, less is known about the early molecular events leading to their formation. Thus, the precise sequence of these events and their relative contribution to the transition from weak initial binding to strong, mature cell-substrate adhesion is still poorly understood. This might be due to the fact that the experimental setup to address these questions has not been available. The aim of the present study was it to study single $α_2β_1$-integrin-collagen interactions and to follow the time-dependent build up of more complex adhesive sites containing $α_2β_1$-integrin. In a first step, an appropriate experimental set-up was chosen based on preliminary experiments.

4.3 Results & Discussion

4.3.1 Preparations

Characterisation of two-dimensional collagen type I matrices (Col)

Col matrices developed in the group of Prof. D. Müller were chosen as substrate for SCFS experiments[301-303]. In agreement with previous work, AFM topographs of Col revealed parallel arrays of collagen fibrils that completely coated the underlying mica surface (Fig. 28A). In AFM topographs recorded at higher resolution (Fig. 28B) Col and pdCol fibrils display the 67 nm D-band characteristic for collagen type I fibrils[304]. This indicated that fibrils of Col/pdCol matrices share structural similarities with fibrils assembled *in vivo*[305, 306]. Most cell attachment studies in the past have been performed with monomeric collagen type I coatings. However, different binding affinities have been reported for $\alpha_2\beta_1$-integrin interactions with monomeric and fibrillar collagen type I[292]. Since *in vivo* collagen type I assembles into fibrils, it is preferable to study integrin mediated adhesion to fibrillar coatings. In previous studies cell adhesion to collagen fibrils that were randomly adsorbed to flat surfaces was non-quantitatively analysed[292]. However, such substrates are not structurally homogenous enough to allow precise SCFS with minimal non-specific interactions. In contrast, the collagen type I matrices shown in Fig. 28 represent a suitable substrate to specifically investigate integrin-mediated adhesion.

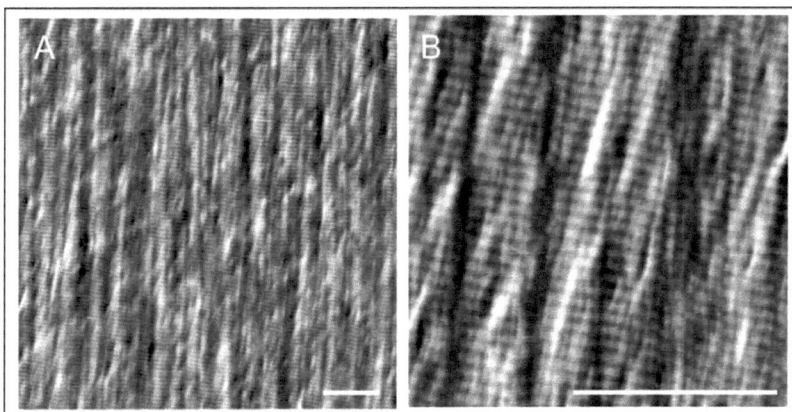

Fig. 28. AFM topographs of two-dimensional collagen I matrices. (A) AFM topographs of a self-assembled collagen matrix. (B) Topograph recorded at higher resolution. Contact-mode AFM topographs were recorded in buffer solution and exhibit vertical scales of 6 nm. Scalebars correspond to 1 µm.

$\alpha_2\beta_1$-integrin expression in Saos-2 and CHO cells

In previous studies binding of $\alpha_2\beta_1$-integrin to collagen type I has been investigated in human osteosarcoma (Saos) and in Chinese hamster ovary (CHO) cells[289]. Both cell lines lack endogenous integrin α_2-subunits. To study the role of $\alpha_2\beta_1$ in mediating adhesion, these cell lines were transfected with human α_2-integrin subunits. Integrin α_2-subunits combine with endogenously expressed β_1-integrin subunits to form functional $\alpha_2\beta_1$-integrin heterodimers[289]. Before these cells were analysed by SCFS in the presented project, α_2-integrin expression levels was confirmed in western blots for both, Saos-2 and CHO cells (Fig. 29).

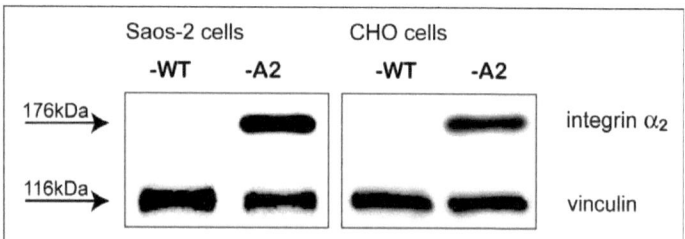

Fig. 29. Integrin α_2 expression in Saos-WT/-A2 and CHO-WT/-A2 cells. Cell lysates of equal protein concentrations were loaded on 7.5 % SDS-polyacrylamide gels and α_2-integrin was detected by western blotting. Vinculin served as a loading controls.

Analyzing the effect of trypsin treatment on cell surface associated $\alpha_2\beta_1$

For SCFS experiments adherent growing Saos-2 and CHO cells had to be detached from the tissue culture dishes. For that purpose trypsin-EDTA solution was used. Trypsin functions by degrading proteins that mediate cell attachment, for instance secreted ECM proteins adsorbing to the surface of tissue culture surfaces and cell adhesion molecules. Furthermore, trypsin-treatment may also degrade adhesion molecules on the cell surface that are later to be analysed in SCFS experiments. To test whether $\alpha_2\beta_1$-integrin cell surface representation was affected by trypsin treatment, CHO-A2 cells were treated for 0 or 10 min with trypsin-EDTA. An extended incubation period was chosen to enhance the eventual degradation effect although prior to SCFS, cells were treated for only 3 min with trypsin. After trypsin-EDTA treatment, total $\alpha_2\beta_1$-integrin concentrations and biotinylated cell surface proteins were compared by western blotting (appendix B13). Western blot analysis did not indicate any effect of typsin treatment, since comparable protein levels (both total and cell surface associated) were found for trypsin-treated and non-trypsin-treated cells. Furthermore, no integrin fragments of lower molecular weight were detected. Moreover, when cell adhesion was quantified by SCFS immediately after trypsin treatment or after a recovery time of 1 h, no differences in adhesion were detected (not shown). Thus, it was concluded that trypsin treatment preceeding SCFS did not affect $\alpha_2\beta_1$-integrin levels.

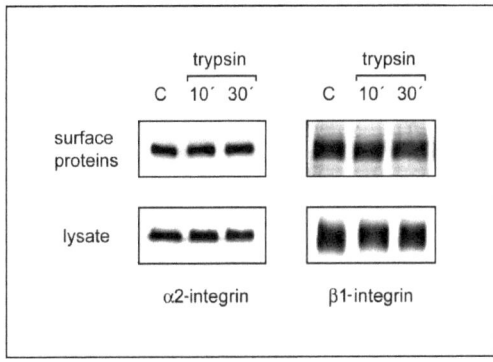

Fig. 30. Effect of trypsin treatment on α_2- and β_1-integrins. CHO-A2 cells were incubated with trypsin-EDTA for 10 min. Lysates of equal protein concentrations were loaded on 7.5 % SDS-PAA gels and α_2- and β_1-integrins were detected in western blots. Cell surface proteins were separated using the EZ-link Sulfo-NHS Biotinylation kit from Pierce.

$\alpha_2\beta_1$-integrin mediated cell attachment and spreading on Col matrices

To compare their spreading morphology, Saos-WT and -A2 cells were seeded on Col. Both cell types could attach and spread on Col, but showed different morphologies after spreading. Whereas Saos-A2 cells were found to spread uni-axially, Saos-WT cells spread multi-axially (Fig. 31 A, B). Apparently $\alpha_2\beta_1$-integrins were responsible for the uni-axial spreading of cells. Previous work of the group showed that $\alpha_2\beta_1$-integrin expressing cells, including Saos-A2 cells, aligned in the fibril direction[307]. It was further demonstrated that cell alignment depended on the ability of cells to exert traction forces onto the collagen fibrils[308]. This was in accordance with studies showing that cells require $\alpha_2\beta_1$ to exert high traction forces[290]. The finding that Saos-WT cells could also attach and spread on Col, suggested that other integrins than $\alpha_2\beta_1$ were involved. Previous studies have identified $\alpha_1\beta_1$-, $\alpha_{10}\beta_1$- and $\alpha_{11}\beta_1$-integrins in Saos-WT cells[292]. Whereas addition of 4 mM EDTA caused rounding of spread cells, addition of 4 mM EGTA had no effect (not shown). EDTA has similar binding affinities for Mg^{2+} and Ca^{2+} ions, whereas EGTA preferentially sequesters Ca^{2+} ions. Thus, spreading of both Saos-WT and -A2 cells was dependent on Mg^{2+} ions in the medium (Fig. 31 C). This is in line with previous studies showing that Mg^{2+} ions are required for collagen binding of integrins[309,275].

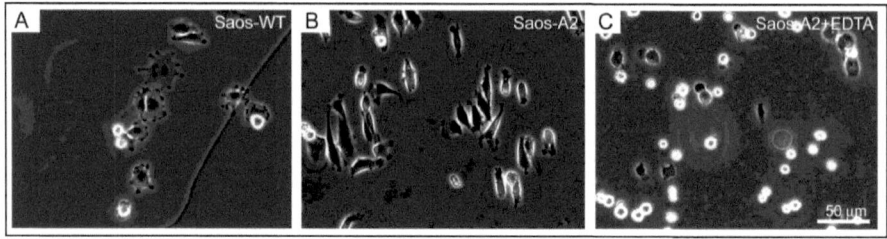

Fig. 31. Spreading of Saos-WT and –A2 cells on Col matrices. Saos-WT *(A)* and Saos-A2 cells in the absence *(B)* or presence *(C)* of 4 mM EDTA. Phase contrast images were taken 1 h after cell seeding.

CHO-WT and -A2 cells seeded onto Col matrices displayed clear differences in attachment and spreading. CHO-A2 cells attached rapidly to Col and spread within 30 min (Fig. 32 B). In contrast, CHO-WT cells did not spread and remained rounded even after prolonged contact (>3 h) with Col (Fig. 32 A). This further indicated that CHO-WT cells lacked endogenous collagen receptors. Fluorescence staining for the α_2-subunit in fully spread CHO-A2 cells showed that α_2-integrin co-localize with classical, elongated focal adhesion contacts (not shown). In absence of Mg^{2+} ions, CHO-A2 cells failed to spread on Col (Fig. 32 C), whereas Ca^{2+}-ion removal had no influence (not shown), similar as observed for Saos-A2 cells. The binding of collagen binding integrins is Mg^{2+}-dependent. In contrast, other collagen receptors, such as discoidin domain receptors 1 and 2 (DDR-1 or –2) that bind in a Ca^{2+}-dependent manner. Consequently, it was concluded that attachment and spreading of CHO-A2 cells on Col was dominated by $\alpha_2\beta_1$-integrins.

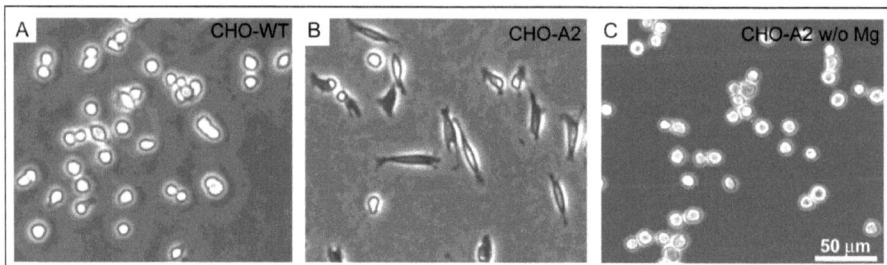

Fig. 32. Spreading of CHO-WT and –A2 cells on Col matrices. CHO-WT *(A)* and CHO-A2 cells in presence *(B)* or absence *(C)* of Mg^{2+}. Phase contrast images were taken 1 h after cell seeding.

Adhesion forces of $\alpha_2\beta_1$-integrin expressing cells measured by SCFS

Next, preliminary SCFS experiments were conducted with Saos-WT and -A2 cells on Col. Individual Saos-WT and -A2 cells were attached to the AFM cantilever and brought for 5 sec in contact with Col matrices. From F-D curves (examples given in Fig. 33 A) cell detachment forces were quantified (2.3.3 and supplementary info B1.2).

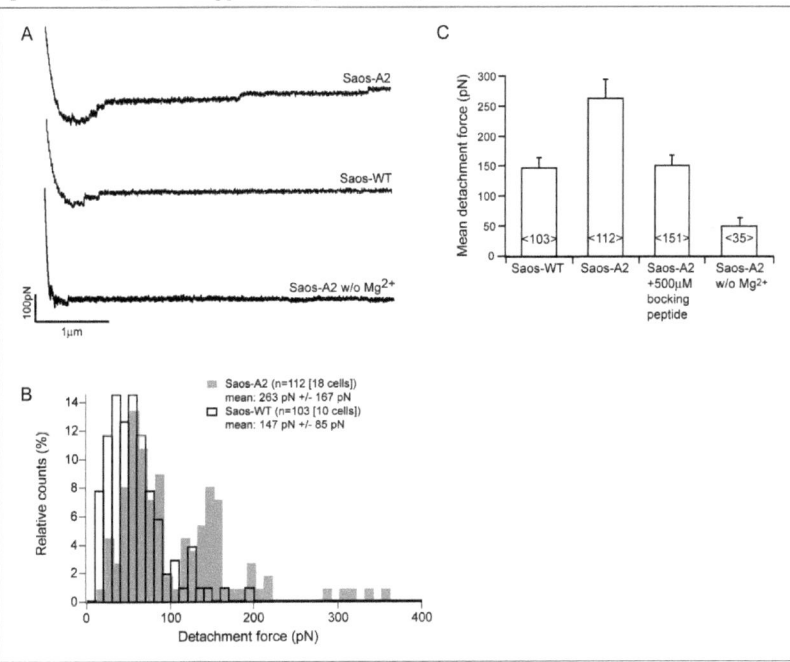

Fig. 33. Quantifying adhesion of Saos-A2 and -WT cells to Col. (A) Representative F-D curves (retrace) recorded for Saos-A2, -WT and -A2 in absence of of Mg^{2+} after 5 sec contact on Col. (B) Distribution of detach-ment forces for Saos-A2 and -WT after 5 sec contact. (C) Mean detachment forces (+/-SEM), recorded for 5 sec contact at different blocking conditions

Significantly increased detachment forces were found for Saos-A2 compared to -WT cells (means: 263 pN versus 147 pN) (Fig. 33 B, C). Detachment forces of Saos-A2 cells were reduced to WT-level in presence of blocking peptide λ229, an effective inhibitor of α_2-integrin-collagen type I interactions (mean: 155 pN). The sequence of the λ229 blocking peptide is deduced from *jararhagin*, a venom from a pit viper (*Bothrops jararaca*)[310].

Addition of 4 mM EDTA significantly reduced adhesion of Saos-A2 (Fig. 33 C) and -WT cells (not shown). Taken together, these results indicated that enhanced adhesion Saos-A2 cells was due to $\alpha_2\beta_1$-integrin binding. These results were consistent with the different spreading behaviour of cells (above). SCFS revealed that Saos-WT cells still could attach to Col matrices, thus, other collagen binding integrins such as $\alpha_1\beta_1$, $\alpha_{10}\beta_1$ or $\alpha_{11}\beta_1$ or other collagen receptors were involved in cell attachment.

SCFS with CHO cells

Next, SCFS was employed to quantitatively measure adhesion of CHO-WT and -A2 cells to Col. F-D curves recorded with CHO-A2 cells (Fig. 34 A) exhibited multiple discrete rupture events, whereas these were rare in CHO-WT F-D curves. In Mg^{2+}-free medium, adhesion of CHO-A2 cells was blocked (Fig. 34 A), but could be restored when Mg^{2+} (0.8 mM) was added to the medium (not shown).

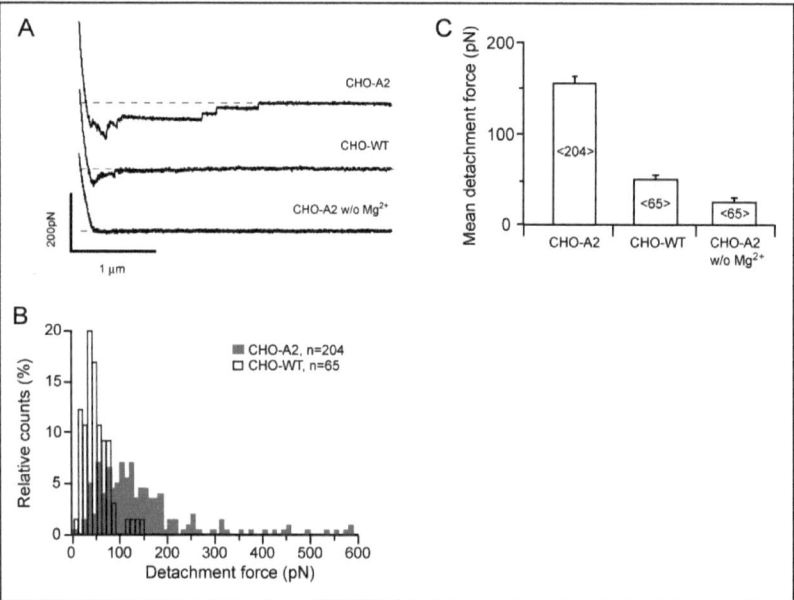

Fig. 34. Quantifying cell adhesion of CHO-WT and -A2 cells to Col. (A) Representative F-D curves (retrace) for CHO-A2 and -WT and -A2 cells in absence of Mg^{2+} after 5 sec contact. (B) Distribution of detachment forces for CHO-A2 and -WT after a contact time of 5 sec. (C) Comparison of mean detachment forces (±SEM) for CHO-WT and -A2 cells with and without Mg^{2+}.

A wide distribution of detachment forces was found for CHO-A2 cell after 5 sec contact; detachment forces between ~10 and 580 pN were detected (Fig. 34 B). In contrast, detachment forces of CHO-WT cells showed less deviations, detachment forces ranged between ~10 and 140 pN. Detachment forces of CHO-A2 cells were significantly higher, 3-fold increased mean detachment forces were found for CHO-A2 (155 pN) compared to CHO-WT cells (49 pN) after 5 sec contact. In the absence of Mg^{2+} CHO-A2 detachment forces were significantly reduced, values were even lower than for CHO-WT cells (Fig. 34 C).

Moreover, cell diameters (Fig. 35 A) of CHO-WT and -A2 cells were compared. Larger cell diameter might increase the contact area that forms between cell and substrate during the measurement and thereby the total number of established integrin-collagen bonds. Similar cell diameters (mean: ~10 µm) were found for CHO-WT and -A2 cells. Furthermore elastic properties of cells were compared by AFM and no differences were found (see 2.3.4, not shown). Thus, it was concluded that increased adhesion of CHO-A2 was not due to different cell size or different mechanical properties of the cells. It was rather assumed that adhesion of CHO-A2 cells to Col was predominantly mediated by α_2-integrins. This was also in agreement with recently performed SCFS experiments in presence of α_2-integrin blocking antibody which reduced adhesion to CHO-WT levels (not shown)[311]. Residual adhesion detected for CHO-WT might be attributed to unspecific interactions or indicate background binding of other non-integrin collagen receptors. Apart from integrins, there are also other collagen binding cell surface proteins, so called non-integrin collagen receptors, for instance DDR-1 and DDR-2[312]. Other collagen receptors include GPVI, cd36, LAIR-1, annexin A5, cd44 and syndecans[273]. However, most of these receptors may predominantly mediate collagen-induced signalling in the cells rather than playing a significant role in mechanically anchoring cells to collagen[273, 312].

The broad distribution of detachment forces detected for CHO-A2 cells, might be attributed to the varying $\alpha_2\beta_1$-integrin expression levels within the analysed CHO-A2 cell population. FACS analysis of CHO-A2 cells fluorescently stained for α_2-integrin revealed a single, log-normal distribution of intensity values (Fig. 35 C, D). This suggested that all CHO-A2 cells exposed α_2-integrins on their cell surfaces. In contrast, the fluorescence signal of CHO-WT cells was comparable to CHO-A2 negative controls, incubated with secondary antibodies only (Fig. 35 C).

Comparing the SCFS results obtained for Saos and CHO cells, it could be concluded that CHO-A2 cells present the better cellular system to specifically analyse $\alpha_2\beta_1$ mediated adhesion, since the background of other collagen binding-receptors was minimal. Thus, all further experiments were conducted with CHO-A2 cells.

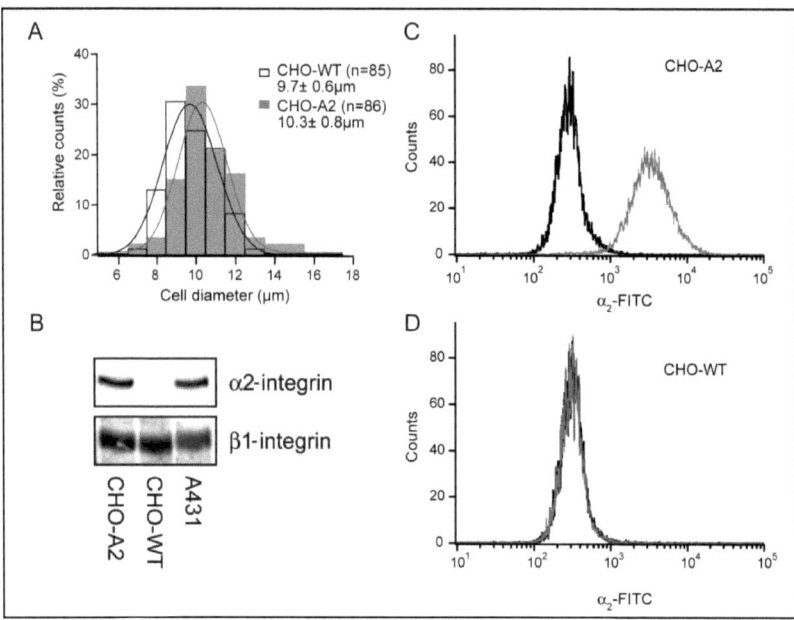

Fig. 35. Analysis of CHO cell diameters and $α_2$-integrin expression (A) Cell diameter distributions for CHO-A2 and -WT cells. (B) Comparing $α_2$-integrin levels in CHO-A2, CHO-WT and a431 cells. Cell lysates of equal protein concentrations were loaded on 7.5% SDS-polyacrylamide gels. $α_2$- and $β_1$-integrins were detected by western blotting. (C)(D) Flow-cytometry of CHO-WT and -A2 cells incubated with antibodies raised against the human $α_2$- integrin subunit (grey). Negative controls (Black) were only treated with FITC-conjugated secondary antibodies.

Overexpression of an exogenous protein might negatively affect normal cell behaviour. Therefore $α_2$-integrin protein concentrations of CHO-A2 were compared to A431 cells that endogenously express $α_2$-integrin[313]. Since $α_2$-integrin levels were similar in CHO-A2 and A431 cells (Fig. 35 B), it was concluded that $α_2$-integrin expression was regulated to levels normally found in cells of epithelial origin.

In the next part, $α_2β_1$-integrin mediated adhesion was analysed in CHO-A2 cells at the single molecule level.

4.3.2 Investigating single $\alpha_2\beta_1$-integrin mediated adhesion events

To investigate $\alpha_2\beta_1$-collagen binding in CHO-A2 cells at the single-molecule level, the SCFS setup was adjusted accordingly (see supplementary info B1.3). The contact area between cells and Col was reduced by minimizing the force by which the cell was brought in contact with the substrate (100-200 pN). Thereby the cell-collagen contact zone was reduced to a minimum. Moreover, a short contact time (50-200msec) was chosen to further reduce the number of interactions between cell and Col. Typically two different types of rupture events can be observed in F-D curves, either rupture events are preceded by a nonlinear force increase ("j", Fig. 36 A) or by a force plateau, "t". j events were interpreted as force-induced unbinding of integrins that were anchored to the cytoskeleton. In contrast, t events were interpreted as the extraction of membrane nanotube (2.3.3, Fig. 20).

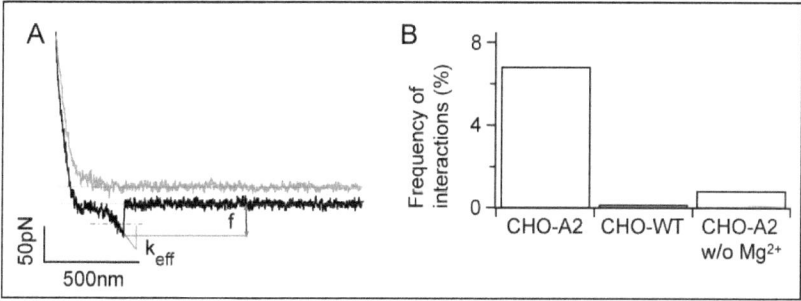

Fig. 36. Detecting single $\alpha_2\beta_1$- integrin Col interactions. (A) F-D curve displaying one single rupture event (f, rupture force; k_{eff}, effective spring constant). (B) The specificity of the $\alpha_2\beta_1$-mediated interaction was confirmed by analyzing the frequency with which interactions occurred in experiments with CHO-A2 cells and controls (i.e. CHO-WT and CHO-A2 cells without Mg^{2+}).

Under the applied conditions approximately 7 % of all CHO-A2 F-D curves displayed j events (Fig. 36 B). j were usually found at pulling distances between 0 and 500 nm. In most F-D curves displaying j, a single rupture event appeared (Fig. 36 A). According to Poisson probability, at a binding frequency of 7 % the probability that a single bond is unbound is 96% (appendix B1.3, Eq. S2). In absence of Mg^{2+} the binding frequency of CHO-A2 cells was decreased to 0.8 % (Fig. 36 B). Moreover, under identical contact conditions, the binding frequency of CHO-WT cells was only 0.1 % (Fig. 36 B). These controls show that j events in CHO-A2 F-D curves corresponded to the unbinding of single $\alpha_2\beta_1$-heterodimers from Col.

Next, dynamic force spectroscopy (DFS) was performed to determine bond-specific parameters, such as the dissociation constant k_{off}, bond lifetime ($1/k_{off}$) and potential barrier width x_u (see 3.3, Fig. 24)[314]. Therefore, single-molecule adhesion measurements were performed at different pulling velocities (Fig. 37 A, B). Pulling velocities between 0.9 to 22 µm/s were chosen, corresponding to effective loading rates in the range of ~200 to 8000 pN/s (appendix B1.3). Rupture forces were mainly normally distributed (normality test, p-value > 0.1) (Fig. 37 A).

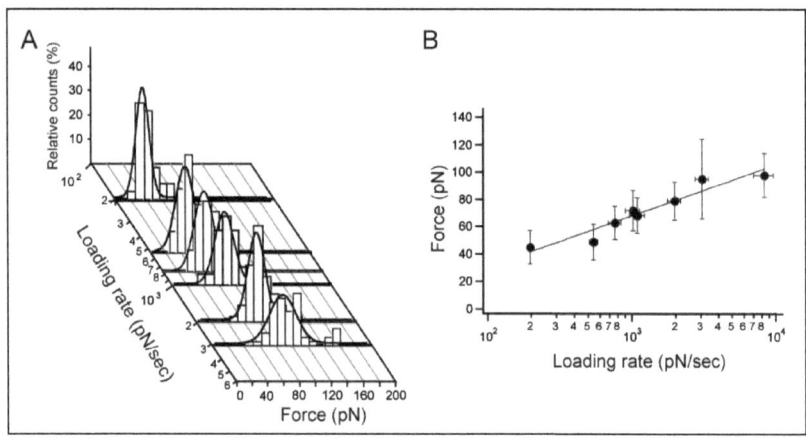

Fig. 37. Analyzing $α_2β_1$-integrin-collagen interactions by dynamic forces spectroscopy. (A) Distribution of rupture forces measured for different loading rates (n>50 adhesion events). (B) Dependence of rupture forces (mean ±SD) on loading rates (mean ±SD). Mean rupture forces were fitted by a line (r^2=0.9) and from the obtained fit parameters (slope, y-intercept), bond parameters k_{off} and barrier width x_u were calculated.

With increasing loading rates the mean rupture forces f^* (here interpreted as binding strength[2]) increased from 38 to 90 pN. Plotting f_m versus the logarithm of the respective loading rates revealed a linear relationship, as predicted by theory (see 3.3)[314]. The constant slope of the binding strength increase suggested the existence of a single energy barrier that was overcome during unbinding. In contrast, the existence of several energy barriers would have been manifested in different regimes, in which the binding strength increases at different slopes (e.g. as shown in Fig. 27, 3.3). Applying Eq. 11 to fit the data points, a dissociation rate k_{off}=1.3 ±1.3 sec^{-1}, corresponding to a lifetime of 0.8 ±0.7 sec, and a barrier width of 2.3 ±0.3 Å were determined for the $α_2β_1$-integrin-collagen type I bond.

SCFS measurements can only detect the unbinding of established integrin-collagen bonds. The *switchblade* model proposes that integrins can either adopt an active, high-affinity or an inactive, low-affinity conformation (see 1.2.3). Accordingly, a ligand can be only bound in the

---

[2] *Since datasets were normally-distributed, there was no significant difference between plotting median, mean or modal.*

high-affinity conformation of the integrin. Therefore, the probed integrin-collagen bonds represent high-affinity binding. Alternatively it has been suggested that ligand binding can also occur in an integrin low-affinity conformation (*deadbold*-model, 1.2.3)[315]. However, this would imply that both low-affinity and high-affinity binding might occur and be detected in parallel. The presented rupture force data exhibit a single force distribution for all analysed loading rates. This suggests the presence of only one (high-affinity) integrin conformation if it may be assumed that force and temporal resolution of the measurements was sufficiently high to detect potential low-affinity conformations.

Moreover, the existence of several high-affinity binding sites within collagen type I was proposed based on solid-phase assays with isolated I-domains[316]. Based on the obtained SCFS data it is hypothesized that there exists either only one single high-affinity binding site for $\alpha_2\beta_1$-integrin or several binding sites with integrin binding strength. In a previous study similar affinities of α_2I-domains towards peptides containing GLOGERGRO and GFOGERGVQ motifs were shown which were assigned as high-affinity binding sites within collagen type I. Binding to both binding sites might result in similar unbinding forces and may therefore not be distinguishable. In the same study another collagen motif (GASGERGPO) was proposed to bind to α_2I, albeit with much lower affinity[316]. If such a third binding site exists, it is possible that this low-affinity binding site could not be detected by the used experimental setup.

The measured binding strength values (f_m=47 pN at 500 pN/sec) are in good agreement with values measured for other integrin-ligand interactions at similar loading rates with AFM. For instance, for $\alpha_5\beta_1$-integrin-FN interactions a binding strength of 55 pN[317], for $\alpha_4\beta_1$-VCAM ~50 pN[318] and for $\alpha_L\beta_2$-ICAM-1/-2 ~50 pN[319] were found.

The calculated dissociation rate of 1.3 s^{-1} compares well with values previously found in AFM-SCFS experiments for $\alpha_4\beta_1$-VCAM-1 interactions (1.1 s^{-1})[320], whereas for $\alpha_5\beta_1$-integrin-FN interactions a smaller dissociation rate was found (0.13 s^{-1})[317]. So far, there has been only one study in which $\alpha_2\beta_1$-integrin-collagen I interactions were analysed performing flow chamber experiments. The authors report a dissociation rate of 0.06 s^{-1}, being about half the value found in the presented AFM data[299]. One possible explanation for the observed differences might be that the binding rates are influenced by the cell type and species in which the respective integrins are analysed. The different results obtained by AFM and flow chambers might be further attributed to the different experimental setups used. Inconsistent results have been reported, for instance, when the lifetime of E-cadherin interactions was analysed by different techniques (flow chambers, BFP and AFM)[321]. Apart from systematic differences that emerge from different analysis methods, the most relevant aspect might be the fact that lifetime is calculated from a dataset recorded at relatively high loading rates. k_{off} is then determined by extrapolating force data detected at high loading rates (see 3.3). The range of loading rates at which data are recorded usually varies among different techniques. Compared to BFP and flow chambers AFM force spectroscopy is limited to relatively high loading rates[321]. Consequently large errors are expected. Moreover, measurements performed at lower loading rates, as in flow chamber experiments, might detect different energy barriers and thereby different values of k_{off}. Another difference between flow chamber and AFM experiments consists in the direction in which bonds are stressed. It has been shown before that the pulling direction can influence the binding strength[322]. This shows that the obtained k_{off} rates must always be discussed in the context of the applied experimental conditions, the technique used and the way bonds are loaded.

Other SCFS experiments conducted with AFM have revealed further activation energy barriers for different integrin-ligand interactions[317, 319, 320, 323] In these studies higher loading rates (> 10000 pN/sec) were applied than in this work although comparable pulling speeds were used. This is explained by differences of the spring constants k_{eff} of the system (membrane, bond, cell cortex, cantilever) (see 3.3). Since cantilever spring constants were similar to the ones used in this work, the differences in k_{eff} might be due to different elastic properties of the cell types used: CHO cells used in this study might be softer compared to small lymphoid cells in previous studies. Since no loading rates >10000 pN/sec could be applied, possible additional activation barriers may not have been detected. Alternatively, the existence of a single energy barrier might reflect differences in the binding mechanism between different integrins and their ligands.

Until now, only j events were discussed. In the F-D curves, occasionally t events were detected (~5 %) which were discriminated from j by the slope preceeding the force step (~0 pN/nm) and the larger distance (> 500 nm) from the surface at which such events occurred. Comparing the magnitude of j and t, major differences were found. Firstly, the forces detected for j at same pulling speeds were significantly higher than those for t events. Moreover, compared with j events (5.90 pNsec/μm), the rupture forces increased slowly with increasing pulling speed (0.83 pNsec/μm) (Fig. 38). During tether formation the underlaying membrane is deformed, thus tether formation depends on the mechanical characteristics of the probed membrane including surface tension and binding rigidity[324-328].

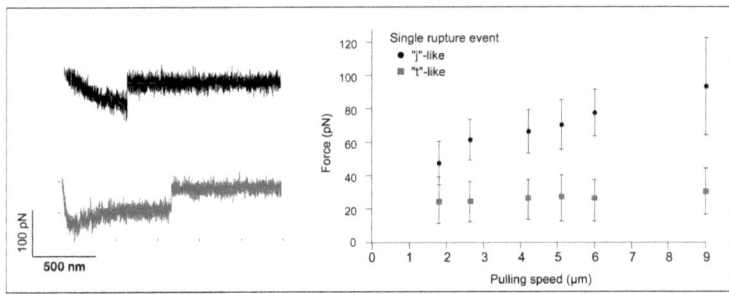

Fig. 38. j- and t- events at different pulling velocities. Mean unbinding force±SD are presented. On the left example F-D retraction curves are shown.

The values for t events that were found in this work (27 pN) are in good agreement with other force spectroscopy studies in which tethers that were pulled out from CHO cells with comparable pulling speeds (28 pN).

The increase of tether forces with increasing pulling speed can be explained by membrane viscosity and friction between membrane and cytoskeleton[327]. The relationship between tether force increase and pulling speed permits the calculation of the effective viscosity (supplementary info B1.3, Eq. S2). From the determined slope of 0.83 pNsec/μm, an effective viscosity of 0.13 pNsec/μm was calculated. This value was smaller compared to a recently reported effective viscosity of 0.33 pNsec/μm determined for tethers pulled by AFM on CHO cells, but very similar to values measured for neutrophil tethers (0.14 pN pNsec/μm)[329].

If tether formation is due to a bond established between a specific receptor and substrate, membrane tethers can also be used to directly calculate bond lifetime[330]. However, in the presented data, no clear differences in the frequency of tether formation between CHO-WT and CHO-A2 cells were seen. Thus, it might be concluded, that most of the formed tethers are contributed to unspecific adhesion events. Indeed, also previous studies performed with living cells concluded that it is not easy and maybe even impossible to distinguish between membrane nanotubes that have formed through specific and unspecific interactions[331, 332].

The results described above show the unbinding of single $\alpha_2\beta_1$-integrin-collagen type I bonds. When a cell is in longer contact with a substrate, the number of interactions is expected to increase with time. In a next step, the time-dependent increase of $\alpha_2\beta_1$-integrin mediated adhesion of CHO-A2 cells to collagen I was therefore investigated.

4.3.3 Dependence of $\alpha_2\beta_1$-mediated adhesion on contact time

Investigating overall cell adhesion

For that purpose, CHO-WT and -A2 cell detachment forces were quantified after contact times between 5 and 600 sec. For all time points detachment forces of CHO-A2 cells were higher compared to CHO-WT cells. This indicated that adhesion of CHO-A2 cells was dominated by $\alpha_2\beta_1$-integrins. During prolonged contact with Col, cells usually maintained their rounded shape on the cantilever. Hence the contact area established between cell and Col did not vary significantly (Fig. 39).

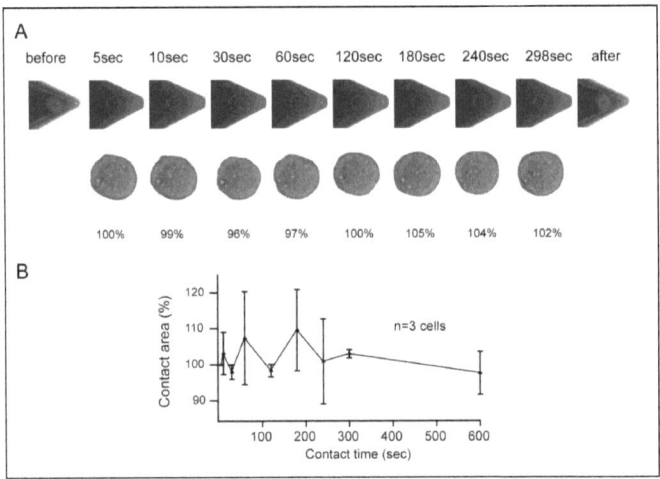

Fig. 39. Monitoring cell shape during contact. (A) Phase contrast images of a CHO-A2 cell on the AFM cantilever during contact with Col. (B) The area of the cell-Col contact zone was measured from the images (A).

Between 0 and 10 sec detachment forces of CHO-WT and -A2 cells continuously increased with time, while no further increase was detected between 10 and 30 sec. The progression of detachment forces in the first 30 sec was well described by an exponential curve (formula given in Fig. 40). Thus, it could be concluded that in the initial attachment period (here 0-10 sec) an increasing number of $\alpha_2\beta_1$-integrin collagen bond had formed, similar as predicted by Eq. 5 (see 3.1).

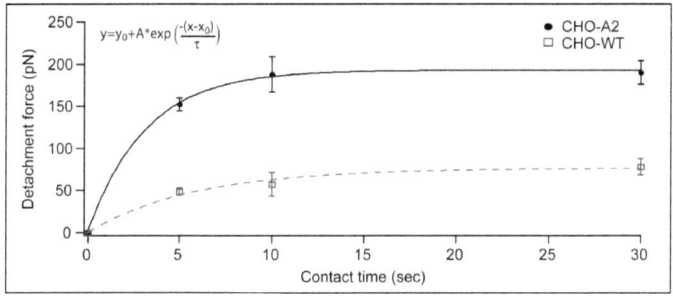

Fig. 40. Detachment forces of CHO-A2 and -WT cells for 0 - 30 sec contact. *Data are presented as mean±SEM. Data were fitted by an exponential fit function (given in figure).*

Recording cell detachment forces over a time course of ten minutes revealed a non-linear build-up of adhesion force. Whereas initially mean CHO-A2 detachment forces grew slowly, they rose quickly after ~60 sec, and reached ~ 5 nN after 180 sec (closed circles, Fig. 41 A). Between 180 and 300 sec, detachment forces did not change significantly. A further increase of detachment forces occurred between 300 and 600 sec.

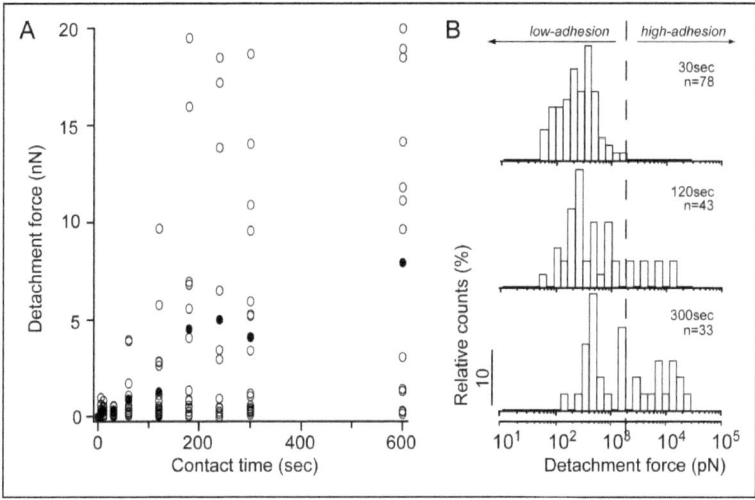

Fig. 41. Dependence of detachment forces on contact time. *(A) CHO-A2 cell detachment forces for 5 and 600 sec contact. Open circles represent detachment forces of individual cells, closed circles corresponding mean detachment forces of all cells tested. (B) Detachment forces for contact times of 30 sec, 120 sec and 300 sec. Note that a half-logarithmic presentation was chosen. The dashed line represents the cut-off force of 2nN chosen to separate low- and high-adhesion cells.*

Since a single-cell technique was used, detachment forces of individual cells could be determined (open circles, Fig. 41 A). Beginning after 60 sec of contact time, cell detachment forces showed considerable variation (~0.5 N to 20 nN).

Monitoring adhesion of individual cells with increasing contact time revealed a sudden adhesion reinforcement at contact times > 60 sec (Fig. 42). For the given example detachment forces initially increased slowly and approached a saturation value. After 3 min of contact the detachment force suddenly rose approximately tenfold (120 sec ~500 pN. 180 sec ~5500 pN). The sudden increase of adhesion suggested that individual cells switched to an elevated adhesive state.

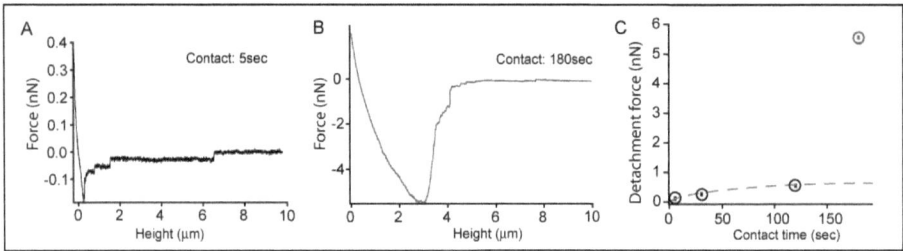

Fig. 42. Example F-D curves displaying increasing detachment forces for prolonged contact. F-D curve recorded for an individual CHO-A2 cell after a contact time of 5 sec (A) and 180 sec (B). (C) Measured detachment forces for contact times between 5 and 180 sec.

This rapid transition of a low-adhesion to a high-adhesion state occurred for different cells at different time points, normally between 60 and 600sec. Two different cell populations were distinguished based on the detected detachment forces (Fig. 41 B, dashed line). Whereas for contact times between 5 and 30 sec detachment forces displayed a uniform distribution and were always below 2 nN (Fig. 41 B), with increasing contact times higher values accumulated although low detachment forces were still present. The fraction of cells exhibiting low detachment forces (0-2 nN) were assigned to a "low-adhesion" group, the group that reinforced adhesion (>2 nN) to a "high-adhesion" group (Fig. 41 B, Fig. 43).

Low-adhesion cells were present over the entire time course. They increased their adhesion during the first 60 sec and subsequently reached a plateau of ~400 sec (Fig. 43 A, inset). In contrast, detachment forces of high-adhesion cells continuously increased between 60 and 600 sec.

This highly-adhesive state was exclusively dependent on α_2-integrins, since detachment forces of CHO-WT cells never exceeded 700 pN during the first 600 sec of contact (not shown).

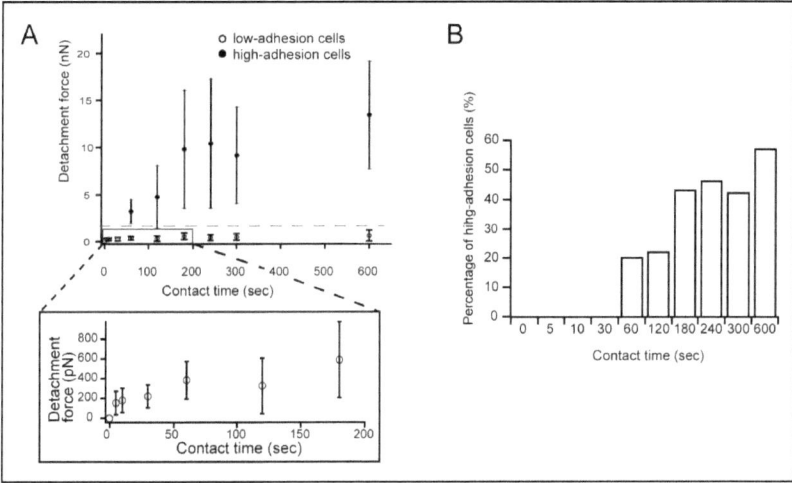

Fig. 43. Separating high- and low- adhesion cells. (A) Detachment forces (mean±SD) of high-adhesion cells (detachment forces > 2 nN) and low-adhesion cells (detachment forces < 2 nN) for different contact times. The dashed line indicates the cut-off detachment force of 2 nN. (B) Time-dependent increase in percentage of high-adhesion cells.

Notably detachment forces detected at short contact (5sec) for low- and high-adhesion cells were similar (not shown). Thus, adhesion reinforcement was not due to increased α_2-expression levels. The ratio of high-adhesion cells increased over time (Fig. 43 B). This indicated that differences in adhesion between both groups could not be explained by the presence of two cell subpopulations with intrinsically distinct adhesive properties. Instead, with increasing contact time cells appeared to switch progressively from a low to a high adhesion state.

What caused the sudden reinforcement of adhesion? It is well established that integrins are clustered in adhesive sites and binding avidity is increased due to cooperative binding[333]. Integrin clustering might be reflected by the magnitude of single rupture events in the F-D curves. These were analysed in the next part.

Analyzing single rupture events in F-D curves

For contact times > 5 sec, CHO-A2 F-D curves usually contained multiple single rupture events (j). These single rupture events represented the smallest detectable force units. They were attributed in previous studies to the unbinding of single or few ligand-receptor bonds[319, 323]. Single rupture events in F-D curves recorded for contact times between 5 and 30 sec showed similar forces (46 ± 16 pN) (mean ± SD) as detected in single-molecule measurements (47 ± 13 pN) at a comparable loading rate (~500 sec/sec) (4.3.2, Fig. 44 A). 100 % of rupture events detected in single molecule experiments (Fig 44 A) were found within a force interval of 0 to 73 pN. Similarly, most single rupture events (90 %) in F-D curves after 5-30 sec of contact lay in this range (Table 3). Also, for low-adhesion cells at increased contact times (120 - 300 sec), the majority of rupture events (75 %) were smaller than 73 pN. In contrast, only 26 % of single rupture events in F-D curves of high-adhesion cells were within the single-integrin unbinding force interval. Magnitudes of single rupture events were significantly increased for high-adhesion cells (159 ± 132 pN)(mean ± SD) (Fig. 44 C). Forces showed a wide distribution and reached values as high as 700 pN, corresponding to a 15-fold force increase above the calculated single-integrin binding strength. Thus, the increase in detachment forces (overall adhesion) of high-adhesion cells coincided with the rise of the single rupture events above the single-integrin level.

Overall adhesive strength ("avidity") of an adhesion complex results from both the total number of receptor-ligand bonds in that complex and the strength ("affinity") of each of these bonds[334]. Receptor clustering may contribute to increased binding avidity because several bonds share the applied force[333]. In contrast, when receptors are uncoupled from each other, receptors experiencing the highest load will become unbound first and the remaining bonds will then rupture in a sequential, zipper-like manner requiring relatively low forces[335]. Furthermore, rebinding events within integrin clusters might account for a force increase of single rupture events.

In addition to cooperative binding, an affinity increase of $\alpha_2\beta_1$-integrin for collagen type I could lead to increased rupture forces (see 1.2.3). However, previous quantitative studies analyzing single integrin-FN rupture forces in the presence of activating antibodies detected only a 0.3-fold force increase (from 60 to 80 pN[317]). Since in the presented data a more than 10-fold enhance force was found for j (reaching values of several hundred pN), it might be assumed that changes in the affinity of $\alpha_2\beta_1$ for collagen are unlikely to account for the strong force increase of single rupture events. Instead, the increase of single rupture events suggested that functional adhesive units containing varying numbers of $\alpha_2\beta_1$-integrins had formed and that cooperative binding of these receptors within small integrin clusters was involved in the time-dependent reinforcement of cell adhesion.

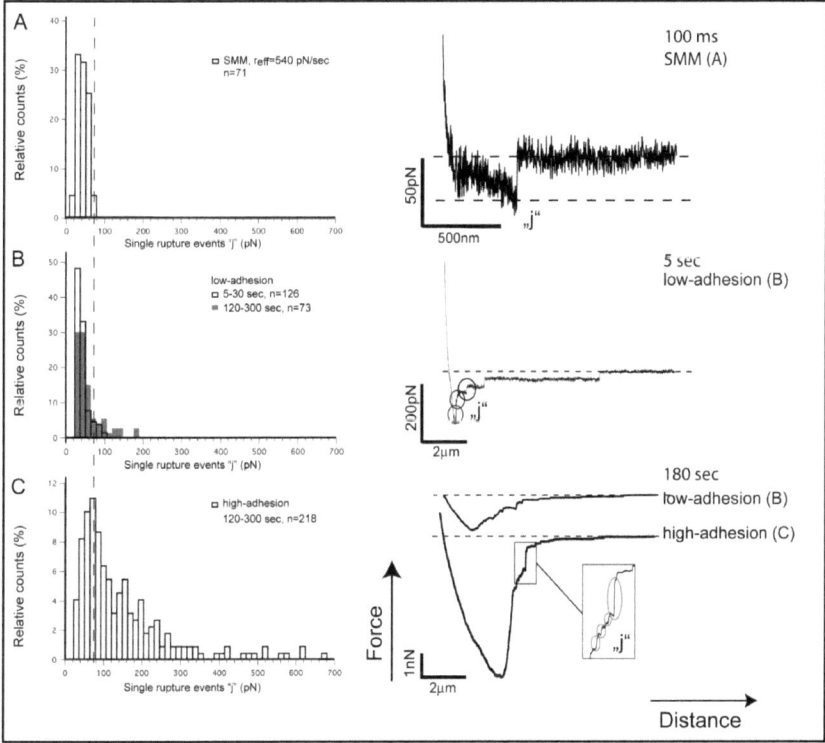

*Fig. 44. **Single rupture events for high- and low-adhesion.** Single rupture events ("j") measured in single molecule experiments (SMM)**(A)**, in low- and **(B)** high-adhesion F-D curves **(C)**. The cut-off force of 73 pN (dashed line) limits the force range encompassing 100 % of the single rupture events **(A)**. On the right example F-D curves are shown.*

Adhesion reinforcement occurred as early between 30 and 60 sec of cell-Col contact. Focal complex formation in cell culture can be observed by fluorescence microscopy 5 - 10 min after replating cells onto an adhesive surface. By actively bringing the cell into contact with the collagen surface during the SCFS experiments, the establishment of specific adhesion might have been accelerated. Furthermore, small integrin clusters that cannot be resolved by conventional fluorescence microscopy might form much earlier than they can be observed optically.

During the rupture of integrin clusters several integrin receptors have to become unbound simultaneously. Understanding the contribution of individual bonds to the rupture force curve during multiple bond rupture is non-trivial. For parallel bond loading, the individual rupture forces have been proposed to be both linearly[336, 337] or non-linearly additive[338]. In force spectroscopy studies on purified receptor/ligand pairs, quantized peaks, corresponding to multiples of the single molecule unbinding forces, were reported[336]. However, the force distribution of the smallest rupture events in high-adhesion cells contained no clear local maxima corresponding to multiples of the single-integrin rupture force of 47 pN. Individual integrin clusters may differ in their linkage to the actin cytoskeleton, resulting in different elastic properties, force loading and ultimately in slightly different rupture forces of the adhesive bridges. Furthermore, due to the geometry of the contact area not all integrin-collagen bonds were probably loaded simultaneously. This might further blur the distribution of single rupture events at increased contact times (see 3.3). Because of the complexity of cooperative integrin-binding in a living cell, the exact number of integrin receptors per cluster could not be measured, but it was roughly estimated that small integrin contacts comprise less than 20 integrin receptors.

Previous studies have reported that myosin II-driven contractility is involved in integrin cluster formation[339]. Thus in the next section the influence of drugs interfering with acto-myosin contractility on the establishment of cooperative integrin binding was tested.

4.3.4 Role of actomyosin contractility for cooperative integrin binding

The effect of two different inhibitors blocking actomyosin contractility on integrin mediated cell adhesion was tested by SCFS. Addition of the myosin inhibitor butandione-2-monoxime (BDM, 20 mM) lead to a reduction of the mean detachment force by more than 50 % at 120 sec contact time and more than 90 % at 300 sec (Fig. 45 A). For the same contact times, the ROCK inhibitor Y-27632 also significantly decreased detachment forces (120 sec: by 63 % and 300 sec: by 91 %). The detachment force decrease in BDM or ROCK inhibitor-treated cells was mirrored by a reduction in the percentage of high-adhesion cells from ~40 % to less than 10 % (Fig. 45 B).

Concomitantly 100% of single rupture events in F-D curves of BDM-treated cells (120-300 sec) ranged in the single-integrin rupture force interval (0-73 pN) in contrast to untreated cells (26%) (Table 3). Similarly, in Y27632-treated cells, about 50 % of single rupture events were <73 pN. These findings indicate that actomyosin contractility is required for establishment of cooperative integrin binding. The Y27632 inhibitor suppressed the establishment of cooperative integrin adhesion less efficiently than BDM. This might be attributed to a more downstream position of the BDM target myosin II in the RhoA-dependent signalling cascade that controls actomyosin contractility

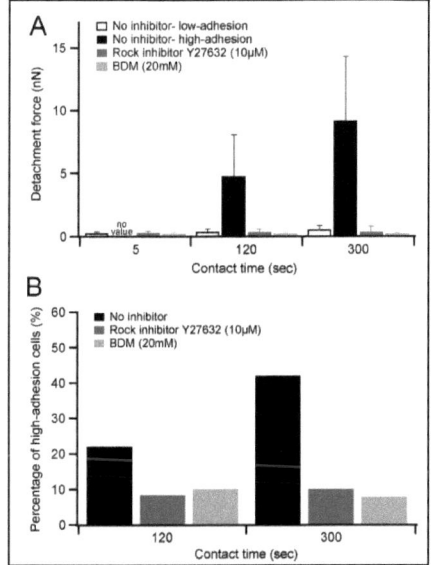

Fig. 45. Influence of inhibitors of actomyosin contractility on cell adhesion. (A) CHO-A2 cell detachment forces (mean+/-SD) in the presence and absence of the ROCK inhibitor Y27632 (10 µM) or the MLCK inhibitor BDM (20 mM). (B) Percentage of high-adhesion cells in the presence and absence of Y27632 and BDM.

The finding that adhesion reinforcement could be prevented by inhibiting actomyosin contractility was in agreement with studies demonstrating the requirement of myosin II-driven contractility to establish and maintain functional focal complexes[339]. Thus, formation of early focal complexes may be myosin II-driven. How may myosin II-activity induce integrin clustering? While unbound integrins are freely diffusive in the membrane plane[340, 341], ligand binding promotes rapid attachment of integrins to actin filaments[342]. It has been suggested in the earlier section that j events detected at contact times of less than a second presented unbinding of cytoskeleton linked integrin in contrast to t events. Myosin II is an effective F-actin crosslinker[343, 344] and myosin-driven bundling and alignment of actin filaments carrying ligand-bound integrin complexes may then lead to the clustering of integrin-cytoskeletal complexes[345, 346]. Furthermore, it has been suggested that myosin-dependent forces exerted on adhesive sites induce conformational changes in associated mechanosensitive proteins. Recently p130Cas has been shown to have a function as mechanosensor involved in transmitting force-dependent signals[347]. Protein-Protein interactions facilitated by conformational changes of such mechanosensory proteins might then promote the growth of focal complexes.

Interval (pN)	SMM (%)	short (%)	non-activated, long (%)	activated, long (%)	w/o inhibitor, long (%)	w/ ROCK inhibitor, long (%)	w/ BDM long (%)
< 73	100	91	75	26	29	56	100
≥ 73	0	9	25	74	71	44	0

Short: contact times from 5 to 30 seconds; Long: contact times from 120 to 300 seconds.

Table 3. Percentage of single rupture events ("j") <73 pN detected for different experimental conditions (inhibitors, contact times). The force interval [0,73 pN] encompassed 100 % of the single-integrin rupture events detected at comparable loading rates during single molecule measurements.

4.3.5 Visualizing paxillin redistribution during SCFS

In the presented work an AFM was used that can be combined with different optical microscopy techniques, for instance conventional fluorescence microscopy, confocal microscopy and total internal reflection microscopy (TIRF). This allows the simultaneous observation of the cell on the cantilever during contact and detachment. By using fluorescently labeled integrins or integrin associated proteins, such as paxillin, it should be possible to follow the formation of focal complexes.

Next, SCFS was conducted with YFP-paxillin expressing CHO-A2 cells. The AFM was mounted on top of a confocal microscope. By adjusting the pinhole settings, the fluorescence signal within an optical slice of <800 nm, including the contact zone between cell and Col, could be visualized during the SCFS experiment. Fig. 46 shows frames of a time-lapse movie recorded during cell contact (15 sec, 60 sec, 300 sec, 600 sec). Diffusive paxillin in the cytoplasm permits visualization of the contact area. In agreement with the previously shown phase contrast images (Fig. 46), the frames indicate that the cell did not spread during contact. Dark areas correspond to the cellular organelles and the nucleus in close proximity to the basal membrane. Interestingly, by the end of the contact period (at 600 sec), bright spots of high paxillin density appear. Whereas non-integrin-associated paxillin is distributed diffusely in the cytoplasma, it is rapidly recruited to nascent focal complexes.

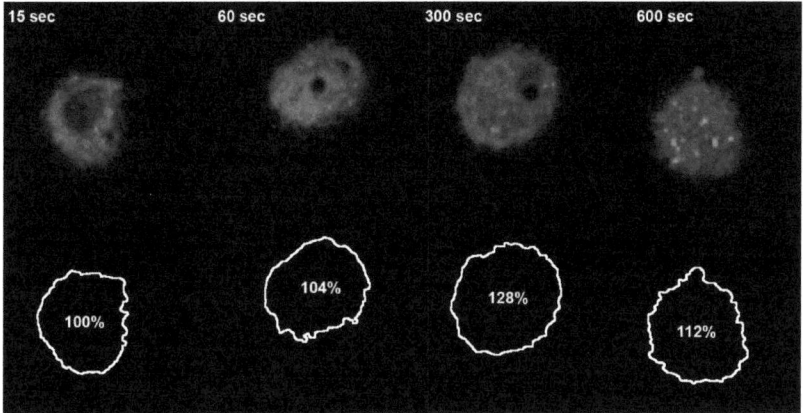

Fig. 46. Confocal microscopy images of the contact zone of a YFP-paxillin expressing CHO-A2 cells and Col during SCFS. The cell was kept in contact with Col for 600 sec. Frames of a itmelapse movie recorded over 610 sec are shown. Below outlines of the contact area, determined by image J, are shown. The contact area was normalized to the area determined in the first frame (15 sec).

Thus, the observed dots might indicate the formation of small focal complexes. However, so far only preliminary results were obtained. Due to the limited temporal resolution of confocal microscopy further experiments should be performed with total internal reflection microscopy (TIRF), a technique that has recently been established in the lab. An increased temporal resolution might enable investigation of the dynamics of formed clusters and might even provide the possibility to correlate single rupture events during cell detachment with optically visualized clusters. Furthermore, such experiments should be performed with a larger number of CHO-A2 and also CHO-WT cells as a control. Correlation of cluster appearance and high-adhesion cells may provide further visual evidence that the switch to high-adhesion is due to the formation of integrin clusters.

4.4 Conclusions

In the presented work $\alpha_2\beta_1$-integrin mediated cell adhesion to collagen type I was quantified within the first 600 sec of contact. Besides characterizing interactions mediated by single integrins, the time-dependent increase in adhesion forces could be precisely monitored.

The generally accepted model describing the time-dependent increase of adhesion strength proposes a two-step process consisting of initial integrin-ligand binding, followed by strengthening of the initial link[348-350]. Adhesion strengthening has been attributed to 1. increase in cell substrate contact area (spreading), 2. receptor recruitment to adhesion sites (clustering) and 3. interactions with cytoskeletal elements[342, 350]. This model has been mainly formulated based on optical microscopy. So far this process has not been investigated with a single-cell technique that allows the detection of single molecule binding events.

The SCFS setup used in this project allowed a detailed description of the time-dependent increase in adhesion. Thereby not only overall adhesion was quantified, but also the contribution of single molecules was investigated. Different regimes of $\alpha_2\beta_1$-integrin mediated adhesion formation could be distinguished: At contact times <0.5 sec adhesion was mediated by a single integrin collagen bonds. Under such conditions bond specific parameters of $\alpha_2\beta_1$-integrin-collagen type I bonds could be determined. Advantageously F-D curves do not only provide information about the integrin-collagen unbinding forces, but also about the mechanical properties of the integrin-cytoskeleton link. The non-linear force increase prior single molecule unbinding suggested that single integrins that mediated initial adhesion were anchored to the cytoskeleton[351]. With increasing contact times adhesion reinforced. The initial increase in overall cell adhesion (0-10sec) could be attributed to an enhanced number of integrin-collagen bonds formed in the contact zone. Apparently, multiple single-integrin interactions enhanced overall adhesion, since number, but not magnitude of single rupture events was increased. Starting at 60 sec contact time cells rapidly reinforced overall adhesion. Concomitantly increased single rupture events were detected that exceeded the level of single-integrin interactions, suggesting the establishment of cooperative integrin binding. These findings are summarized in Fig. 47. Conducting SCFS in the presence of drugs interfering with actomyosin contractility further revealed the involvement of actomyosin contractility in establishment of cooperative integrin binding. This might indicate a role of myosin II in crosslinking cytoskeleton-associated integrins[343, 344]. Taken together, the kinetics of integrin-mediated adhesion could be precisely described, and insights into the mechanisms required for cooperative integrin binding were obtained. Currently, there are no other techniques that allow such insights in adhesion kinetics. This is owed to the high force resolution and precise control of the contact zone and contact time in AFM SCFS experiments.

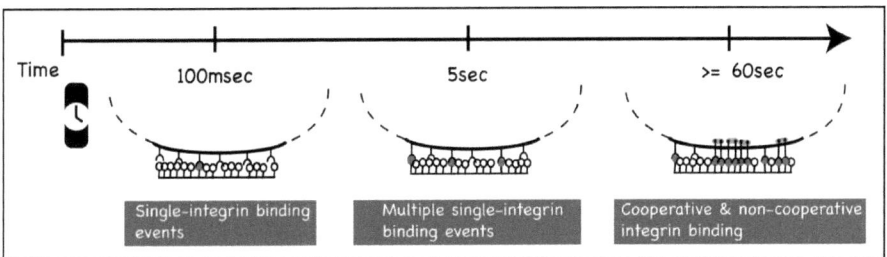

Fig.47. Overview about the sequential build-up of $\alpha_2\beta_1$-integrin-collagen type I bonds. Initially adhesion is dominated by single molecule recognition events. Within the second-range, multiple integrin-interactions occur. At contact times >1 min cooperative integrin binding can be observed concomitant with reinforcement of overall cell adhesion.

It has been suggested that integrins cover the largest range of binding forces compared to other adhesion molecules, for instance selectins and cadherins[351]. Indeed, requirements on integrin binding in tissues are quite diverse. High bond dissociation rates are suggested to permit the fast remodelling of integrin-matrix interactions which is indispensable for cell migration[346]. In other situations, cell-ECM links have to be reinforced quickly to balance raising external forces[351].

It might be expected that varying off-rates between different types of adhesion molecules reflect different requirements of the specific integrin-ligand pairs. Surprisingly, similar binding strengths (~50 pN) (at same loading rates) are found for different types of adhesion molecules, for instance different types of cadherins [352] and other integrins (e.g. $\alpha_5\beta_1$, $\alpha_4\beta_1$, $\alpha_L\beta_2$, see above). This suggests that the large force range covered by integrin binding is explained by the manifold regulatory mechanisms that control integrin binding, e.g. integrin affinity regulation (number of functionally active integrins) and cooperative integrin binding. In the presented project adhesion reinforcement was predominantly caused by cooperative integrin binding. Thus it may be hypothesized that cooperative integrin binding plays a dominant role in regulating integrin mediated adhesion, whereas affinity regulation might be more important in cell types in which integrins are not constitutively active, for instance in platelets or leukocytes.

Chapter 5

Effects of cryptic integrin binding site exposure in collagen type I on osteoblast adhesion and matrix mineralisation

5.1 Abstract

Collagen type I contains cryptic binding sites for integrins that can be exposed upon thermal denaturation. Aim of this project was it to quantitatively study the effect of these binding sites on pre-osteoblast adhesion and subsequent differention. Pre-osteoblast adhesion to native (Col) and partially-denatured (pdCol) collagen I was compared by SCFS. During early stages of cell-attachment (0-180 secs) cells showed significantly enhanced adhesion to pdCol. Adhesion to pdCol was reduced by soluble RGD (Arg-Gly-Asp)-peptide indicating the exposure of RGD-motifs in pdCol. Accordingly, experiments performed in presence of integrin blocking antibodies revealed that $\alpha_5\beta_1$- and α_v-integrins mediated cell adhesion to pdCol, but no to Col. Pre-osteoblasts seeded on pdCol increased their focal adhesion kinase tyrosine-phosphorylation level compared to Col. Concomitantly enhanced spreading and motility were observed on pdCol. Cells cultured over six weeks on Col and pdCol showed no differences in proliferation. However, pdCol matrix mineralisation was more pronounced at all analysed time points. The presented data suggest that partially-denaturing collagen I exposes RGD-motifs that trigger differential integrin binding and signalling and thereby stimulate osteoblastic cell differentiation. The finding of this project might open new perspectives for the development of optimized tissue culture substrate.

5.2 Introduction

RGD-motifs are common integrin recognition sites within several ECM proteins for instance osteopontin, FN or vitronectin[353-355]. Also collagens contain several RGD- motifs. In collagen type I, for instance, the α_1 and α_2 polypeptide chains forming the collagen type I triple-helices, exhibit two and four RGD-motifs, respectively[356]. In the native collagen conformation these cryptic RGD-motifs are structurally not acessible to integrin binding. Therefore, cells adhere to native collagens through RGD-independent integrin interactions[357-360] (see 4.2). This rises the question if there is a particular role of such cryptic RGD-motifs in collagen type I and if there are situation in which these become exposed.

Fourier transform infrared micro-spectorscopy (FTIRS) studies demonstrated that thermal and proteolytic denaturation of collagen I leads to unwinding of the rigid triple-helical structure[361-363]. Since RGD-dependent cell attachment of cells to thermally denatured collagen I was observed in several studies[358, 364, 365], it was proposed that denaturation coming along with unwinding of the triple-helix exposes the hidden RGD-motifs[361, 364, 366] (Fig. 48).

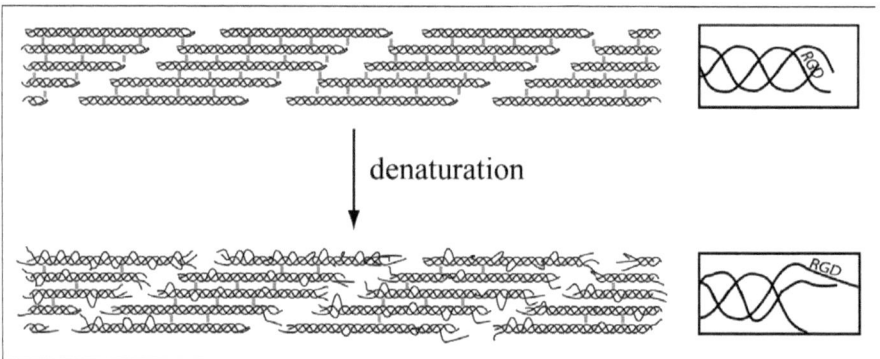

Fig. 48. Sketch illustrating the effect of thermal denaturation on collagen type I structure. Thermal denaturation leads to partial unwinding of the collagen triple-helices and exposure of cryptic RGD motifs. Grey bars outline intermolecular interactions stabilizing collagen fibril structure.

There are at least seven different integrin heterodimers that bind their ligands in a RGD-dependent manner[367]. However, so far, there is controversy about the integrin heterodimers that recognize denatured collagen type I; whereas some authors identified integrin $\alpha_V\beta_3$ as major integrin heterodimer[364], others reported that predominantly integrins $\alpha_V\beta_1$ and $\alpha_5\beta_1$ mediated cell adhesion to denatured collagen type I[358]. The contradictory results might be contributed to the different integrin expression patterns and levels of the studied cell types. Furthermore, the protocols used to produce gelatin, i.e. duration of heat treatment and denaturation temperature, varied among different studies. This might significantly influence the structure of denatured collagen: harsh thermal denaturation (90 °C), for instance, has been shown to cause melting of triple-helical strands[363]. Another reason for the inconsistent results might be that exclusively non-quantitative adhesion assays were applied. In the presented work AFM-based SCFS as a quantitative and sensitive approach has been applied.

How relevant is denatured collagen for cells *in vivo*? Collagen matrix degradation occurs in many normal and pathological situations. For instance, the ECM in developing tissues, but also in adult animals is subjected to a continuous remodeling process which is essential to ensure tissue morphogenesis and homeostasis[368-372]. Collagen remodelling occurs as a consequence of controlled biodegradation and *de novo* synthesis and achieves a state of equilibrium between native and denatured collagen[373]. In certain pathological situations this balance is disturbed, and massive collagen degradation occur, for instance in osteoarthritis or in skeletal metastasis[373]. It has been postulated that exposure of so-called *matricryptic* sites in ECM proteins represents a general mechanism by which specific signals are conferred to cells[364, 366]. RGD motifs were proposed to be common *matricryptic* sites, amongst others (e.g. laminins) also in collagen. Exposure of RGD-motifs might be of physiological relevance in situations in which tissue collagen is structurally altered, for instance during normal tissue remodeling to achieve homeostasis or after tissue injury. The produced signals might regulate cellular behaviour within the injury site and thereby assist tissue repair[366].

Whereas collagen type I fibrils have been largely applied for coating and patterning of non-biological surfaces to enhance biocompatibility[374-376](see 1.1.1), so far the use of denatured collagen has not been exploited. Respective studies have been performed by Kaplan's group[377, 378]. It was found that mesenchymal stem cells exhibited enhanced osteogenic differentiation capacity during *in vitro* expansion on denatured collagen compared to native collagen[377, 379]. Based on these findings it was suggested that RGD-dependent integrin interactions with denatured collagen can produce cellular signals retaining cellular functions that are normally lost upon extensive cell passage. This might be an interesting aspect for the development of optimised substrates for the *ex vivo* expansion of stem cells in regenerative medicine.

In the presented work pre-osteoblast cell adhesion to native and denatured collagen I was quantitatively compared. Further the contribution of specific adhesion receptors in mediating adhesion has been studied. To test the effect of collagen type I matricryptic sites, moreover cell proliferation and differentiation were compared among native and denatured collagen type I matrices in six weeks´ cultures.

5.3 Results & discussion

5.3.1 Characterization of Col and pdCol matrices

AFM imaging of native and thermally denatured collagen type I matrices

In the presented project collagen type I matrices (Col) described in chapter 4 were used as adhesive substrates[380]. In a first step the topography of Col before and after thermal denaturation (1 h, 50 °C, appendix B2) was compared by AFM (appendix B3). AFM topographs revealed that the fibrillar assembly typically observed for Col (Fig. 49 A) was maintained after heating. Matrices were thereupon referred to as partially-denatured collagen type I matrices (pdCol). Topographs recorded at higher resolution (Fig. 49 A, B; insets) confirmed that both, Col and pdCol fibrils, displayed the 67 nm D-band inherent to collagen type I fibrils assembled *in vivo*[381]. Since no major structural differences between Col and pdCol matrices were observed, it was concluded that thermal denaturation did not affect the macromolecular collagen type I assembly.

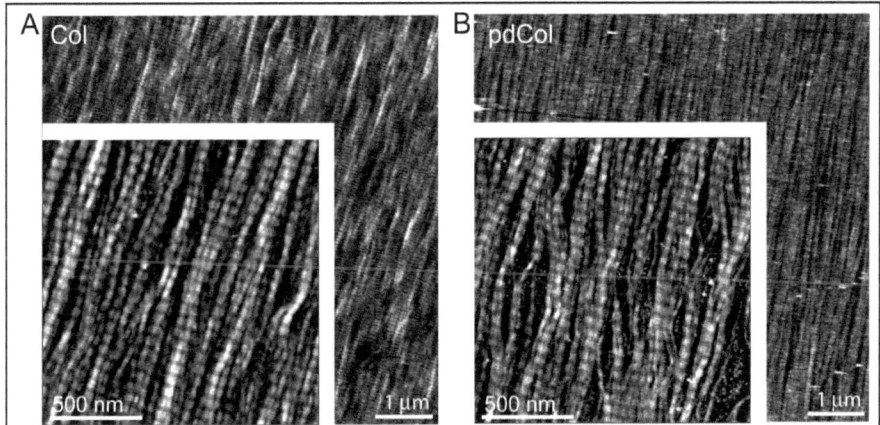

Fig. 49. Characterization of two-dimensional collagen type I matrices. AFM topographs of Col (A) and pdCol (B). Insets show Col/pdCol at higher resolution. Topographs were recorded in buffer solution and exhibit vertical scales of 6nm.

Immunofluorescence staining of Col and pdCol

To analyse thermally denatured Col matrices at the molecular level, the binding affinity of a monoclonal collagen I antibody (mAB) to Col and pdCol was compared by immunofluorescence microscopy (appendix B5). mAB-stained Col showed a strong immunofluorescence compared to pdCol exhibiting no staining (Fig. 50 A, B). The high binding affinity of this particular mAB had been reported to depend on the triple-helical structure of native collagen[382]. This affinity was lost in case of pdCol suggesting an altered structural integrity of thermally treated collagen fibrils. Labelling Col and pdCol with a polyclonal collagen antibody (pAB) resulted in uniform immunofluorescent staining in both cases (Fig. 50 A), confirming- in addition to AFM topographs- that both matrices homogenously covered the mica.

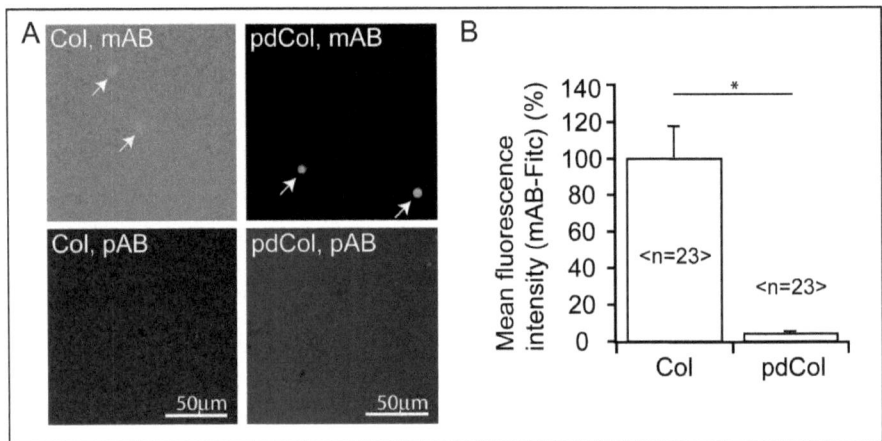

Fig. 50. Characterisation of Col and pdCol by immunofluorescence microscopy. (A) Confocal microscopy of immuno-labelled Col/pdCol. Matrices were stained with monoclonal antibodies (mAB, green) and polyclonal antibodies against collagen type I (pAB, red). FITC-conjugated beads (arrows) were added as reference. (B) Mean fluorescence intensities ± standard deviations (SD) measured for mAB-stained Col/pdCol. Brackets give numbers of analysed images.

Comparison of the mechanical properties of Col and pdCol

Subsequently the mechanical properties of Col and pdCol were compared by AFM (appendix B4). 2x2 µm² sized sections of Col and pdCol were scanned applying increasing forces to the AFM tip (Fig. 51A, B). The manipulated areas were re-imaged at minimal forces ≈ 50 pN) to evaluate the structural changes. Forces of 4.3 nN structurally altered Col matrices, whereas on pdCol major structural changes begun at 2.7 nN (Fig. 51D). Structural rearrangements observed on Col were less severe; Col fibrils stayed mainly intact, but were occasionally shifted by the AFM tip. In contrast, pdCol fibrils were frequently disjointed exposing the underlying mica (Fig. 51 A, B; arrows). This experiment demonstrated that Col matrices were mechanically more resistant. This might be explained by the structural changes that occurred within Col during thermal denaturation.

Unwinding of the rigid triple-helical structure might have reduced or weakened intermolecular interactions that stabilized Col fibrils (see Fig. 48).

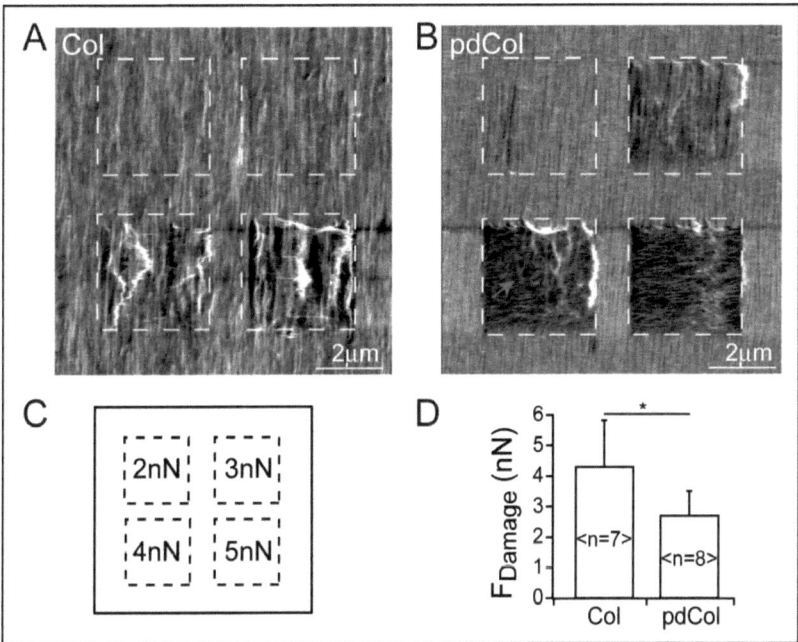

Fig. 51. Characterizing mechanical properties of Col and pdCol. (A, B) AFM topographs of Col/pdCol matrices that had been scanned under increasing forces (see C). (D) Mean forces (±SD) at which the scanning AFM tip initially induced structural deformations (F_{damage}). In brackets numbers of analysed samples are given.

5.3.2 Analyzing cellular interactions with Col and pdCol

Analyzing cell spreading and migration on Col and pdCol

Next spreading and migration of pre-osteoblastic MC3T3-E1 cells seeded on Col and pdCol were investigated (Fig. 52 A). Time-lapse movies showed that cells spread and elongated along collagen fibrils on both types of matrices, similarly as described for Col in earlier work of the Müller group[383]. There were no gross differences in spreading morphology on both matrices. However, cells spread significantly faster on pdCol than on Col (Fig. 52 A, B). On pdCol 50 % of cells had spread after ~14 min, on Col only after ~26 min. This suggests that cells established more adhesive interactions to pdCol.

Fluorescence images revealed considerable reorganisation of Col and pdCol fibrils. Concomitantly with enhanced spreading on pdCol, remodeling of pdCol matrices started earlier on pdCol (Fig. 52 C). Increased matrix remodeling might be attributed to enhanced adhesive interactions of MC3T3-E1 cells with pdCol or to the decreased mechanical stability of pdCol. Col has been demonstrated in above experiments to be mechanically more resistant to lateral scratching forces than pdCol. Thus, Col fibrils might have shown more resistance to cellular traction forces compared to pdCol.

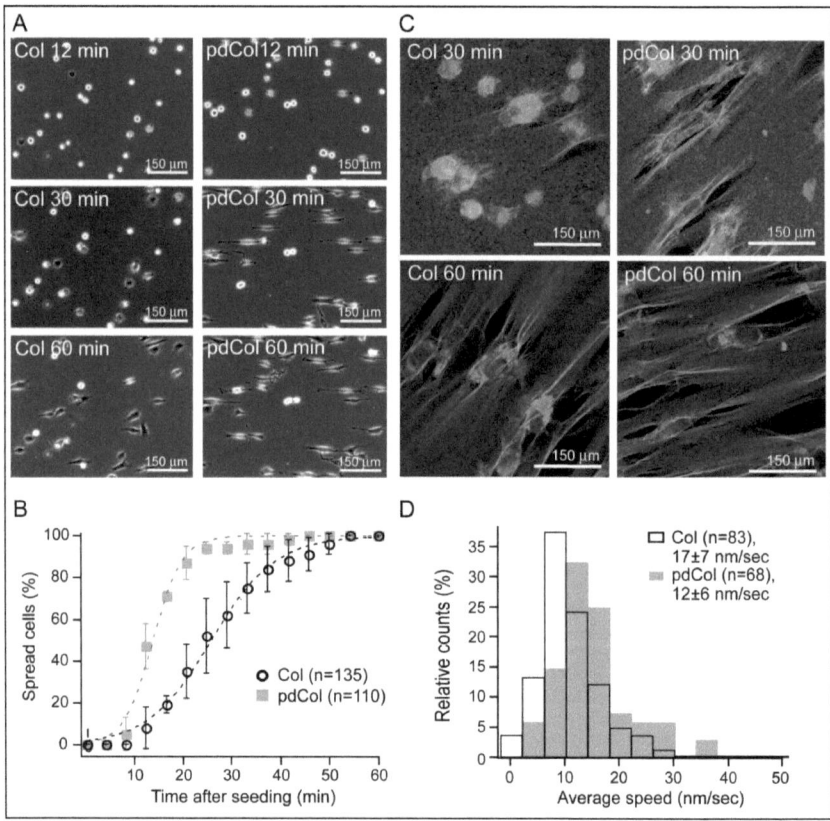

Fig. 52. Cellular behaviour on Col/pdCol matrices. (A) Frames taken from time-lapse movies of MC3T3-E1 cells on Col/pdCol. Images correspond to time points of 12, 30 and 60 min after cell seeding. (B) Cell spreading (mean±SD) within the first 60 min after seeding. Sigmoidal curves revealed the time after which 50 % cells have spread. In brackets numbers of analysed cells are given. (C) Deformation of Col/pdCol matrices by MC3T3-E1 cells (collagen-green, actin-red, nuclei-blue). (D) Migration speed of individual MC3T3-E1 cells seeded on Col/pdCol. Mean speed±SD are annotated above histograms.

Furthermore, the motility of MC3T3-E1 cells seeded on Col and pdCol was analysed. The relative displacement of single MC3T3-E1 cells was tracked in time-lapse movies. The average migration speed was enhanced by 42 % on pdCol (Fig. 52 D). Above observed differences in spreading, remodelling and migration on Col and pdCol suggested that pre-osteoblasts interacted differently with both matrices. To quantify the adhesive interactions with Col and pdCol, in a next step SCFS was applied.

Comparing MC3T3-E1 cell adhesion to Col and pdCol by SCFS

F-D curves were recorded on Col and pdCol using contact times of 5, 30 and 180 sec (Fig. 53 A).

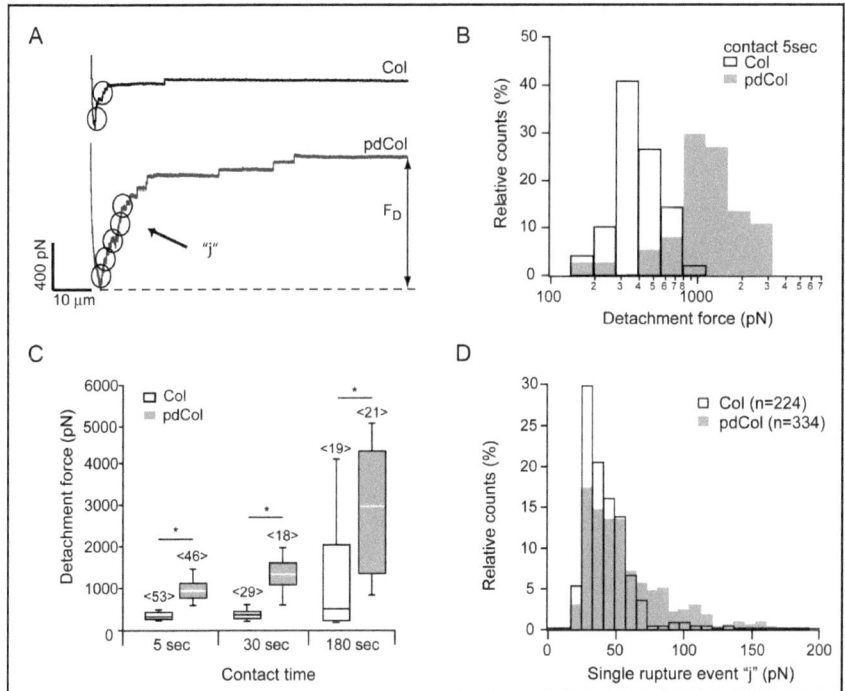

Fig. 53. Quantification of cell adhesion to Col/pdCol matrices by SCFS. (A) Detachment forces of MC3T3-E1 cells adhering to Col/pdCol matrices. (B) Quantification of single rupture forces ("j", (b)) recorded after a contact time of 5 sec. Brackets annotate numbers of events analysed. Data is represented as Box-Whisker plots, white lines within boxes denote medians, and <n> total numbers of analysed F-D curves.

Pre-osteoblastic cells showed significantly increased median detachment forces (F_D, 2.3.3) on pdCol (Fig. 53 B, C). A broader distribution of detachment forces was found on pdCol compared to Col (Fig. 57 B). After a contact time of 5 sec, a threefold increased detachment force was detected on pdCol (1025 pN) compared to Col (334 pN). This difference became more prominent after enhanced contact times (2856 pN *vs.* 522 pN after 180 sec). This quantitatively showed that overall adhesion was enhanced, either due to a higher number of adhesion receptors that were binding, or due to an increased binding strength of individual receptors. To enter into this question, single rupture events ("j") in F-D curves were analysed (Fig. 53 A, D, see 2.3.3). These can be correlated with the unbinding of single or few cell adhesion molecules[384, 385].

A significantly higher number of j per F-D curve was detected on pdCol (7.8±5.5 *vs.* 3.4±2.6). The magnitude of j was also increased on pdCol (49 pN *vs.* 37 pN, p<0.0001) (Fig. 53 D). The differences of j let suggest that distinct adhesion molecules were binding to Col and pdCol.

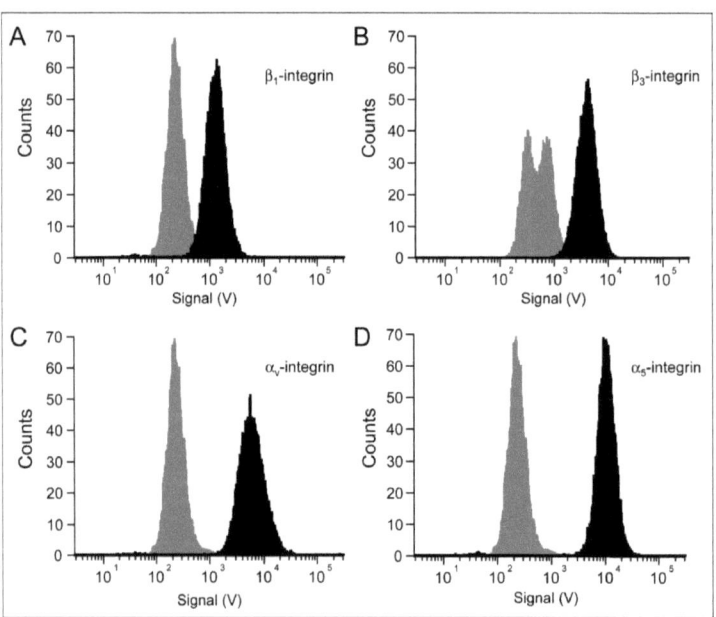

Fig. 54. Flow cytometry analysis of cell surface associated integrins. *β_1-, β_3-, α_v- and $\alpha_5\beta_1$-integrins were immuno-fluorescently labelled and analysed by flow cytometry (black). Negative controls are shown in grey.*

Next, the role of β_1-, α_V-, $\alpha_5\beta_1$- and β_3-integrins in mediating adhesion to Col was studied, because some of them were described to mediate adhesion to denatured collagens[358, 364, 386, 387]. Firstly presentation of these integrin heterodimers on MC3T3-E1 cells was confirmed by flow cytometry (appendix B12, Fig. 54).

The contribution of these integrins to overall cell adhesion was quantitatively investigated in presence of integrin blocking antibodies (β_1, β_3, α_V, $\alpha_5\beta_1$) or linear RGD-peptide. Blocking of β_1-integrin reduced adhesion to Col by 20 % (p=0.006) (Fig. 55 A). In contrast, the addition of the other blocking antibodies and RGD-peptide led to enhanced adhesion to Col (α_V: +20 %, $\alpha_5\beta_1$: +55 %, β_3: +40 %, RGD: +10 %). Integrin crosstalk might have caused the observed effect; blocking of $\alpha_5\beta_1$-integrins and integrins comprising α_V and β_3 subunits may have activated collagen-binding integrins. A similar crosstalk has been reported for $\alpha_5\beta_1$- and $\alpha_2\beta_1$-integrins *via* a protein kinase C dependent mechanism[388]. The fact that adhesion to collagen could not be completely abolished in presence of β_1-integrin blocking antibody suggested that either other collagen binding receptors or unspecific adhesion events accounted for the residual binding. Alternatively, β_1-integrins might not have been efficiently blocked by the antibody.

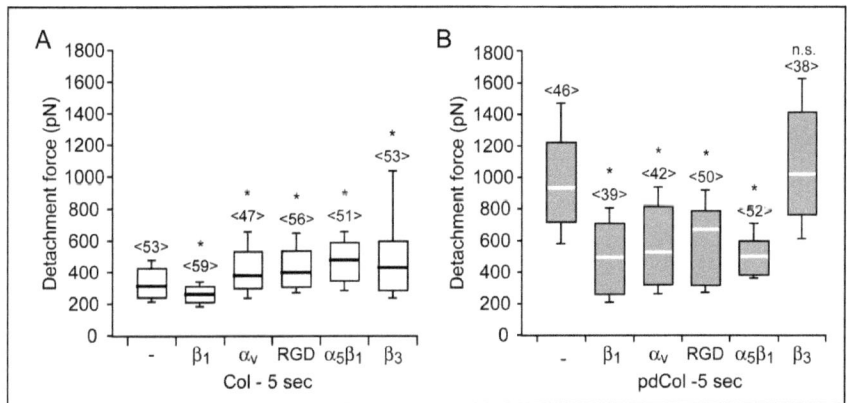

Fig. 55. Quantifying the effect of integrin blocking on cell adhesion to Col (A) and pdCol (B). Detachment forces of MC3T3-E1 cells (5 sec contact time) that had been pre-incubated for 30 min with different antibodies blocking β_1-, α_V-, $\alpha_5\beta_1$- or β_3-integrins or RGD peptides. Data is represented as Box-Whisker plots, lines within boxes denote medians, and <n> total numbers of analysed F-D curves.

In contrast, MC3T3-E1 cell adhesion to pdCol was decreased by antibodies blocking α_V-, $\alpha_5\beta_1$-, and β_1-integrins or by RGD peptide (α_V: -45 %, $\alpha_5\beta_1$: -45 %, β_1: -40 %, β_3: -50 %, RGD: -

30 %). Within the initial attachment period (5-180 sec, only 5 sec shown) blocking of β_3-integrin had no significant effect on adhesion (p=0.16)(Fig. 55 B). This indicated that $\alpha_5\beta_1$- and α_V-integrins probably as part of the integrin heterodimers $\alpha_V\beta_1$ and/or $\alpha_V\beta_5$, could bind to pdCol, but not to Col matrices. These three integrin heterodimers are known to bind in a RGD-dependent manner to fibronectin ($\alpha_5\beta_1$, $\alpha_V\beta_1$), osteopontin ($\alpha_5\beta_1$, $\alpha_V\beta_1$, $\alpha_V\beta_5$) and vitronectin ($\alpha_V\beta_1$, $\alpha_V\beta_5$)[367]. Similarly pre-osteoblast adhesion to pdCol was not completely reduced in presence of β_1-blocking antibody. This might indicate a role for other integrin heterodimers as $\alpha_V\beta_5$ in recognizing pdCol. Further, incomplete blocking might be attributed to unspecific binding, inefficient antibody blocking or other non-integrin collagen receptors. Previous studies suggested that denatured collagen is not recognized by receptors for native collagen including collagen binding integrins as $\alpha_2\beta_1$-integrin or non-integrin receptors, such as discoidin domain receptors (DDR)[361].

Above results indicate that enhanced pre-osteoblast adhesion to pdCol was mediated by integrins that recognized RGD-motifs. These findings are in line with previous studies analyzing cell adhesion to native or denatured collagens[358, 364, 365, 386, 387]. However, there has been controversy about the involved heterodimers. Whereas some groups suggested that $\alpha_V\beta_3$-, but not $\alpha_5\beta_1$-integrin is the major cell adhesion molecule binding to denatured collagens[364, 365, 387], others reported a role of $\alpha_3\beta_1$, $\alpha_V\beta_1$ and $\alpha_5\beta_1$[358, 386]. Although $\alpha_V\beta_3$-integrin was present on MC3T3-E1 cells (Fig. 54), this integrin heterodimer was apparently not binding to pdCol, since β_3-blocking antibody had no effect on cell adhesion to pdCol. These contradictory results may be given by the different experimental setups used to characterize cell adhesion. Some studies have used collagen I[358, 364, 365], others focused on collagen II[386] and VI[387]. The obtained results might be further dependent on the cell type studied, since integrin expression pattern and levels vary among different cell types. Moreover, different protocols were used for collagen denaturation and the applied adhesion assays hardly allowed to obtain quantitative data on cell adhesion. Previous studies either probed adhesion of isolated molecules[364] or applied washing assays. Whereas SCFS analyses the initial cell adhesion (5 - 180 sec), washing assays are restricted to longer attachment periods (> 30 min). It is possible that at prolonged contact periods, different integrin heterodimers might come into play. Furthermore, upon longer contact, additional ECM proteins might be secreted and perturb the measurements. Moreover, results provided by washing assays do not necessarily directly correlate with cell adhesion, since other effects, for instance cell spreading morphology, come into play (see 2.1.1).

5.3.3 Analyzing effects of Col and pdCol on cell growth and differentiation

FAK phosphorylation at tyr 397 in MC3T3-E1 cells seeded on Col and pdCol

In a next step the consequences of different integrin-binding on focal adhesion kinase (FAK) phosphorylation was investigated (appendix B8). FAK localizes to sites of clustered integrins and becomes phosphorylated; one of the first FAK phosphorylation sites upon integrin engagement is residue tyr397.

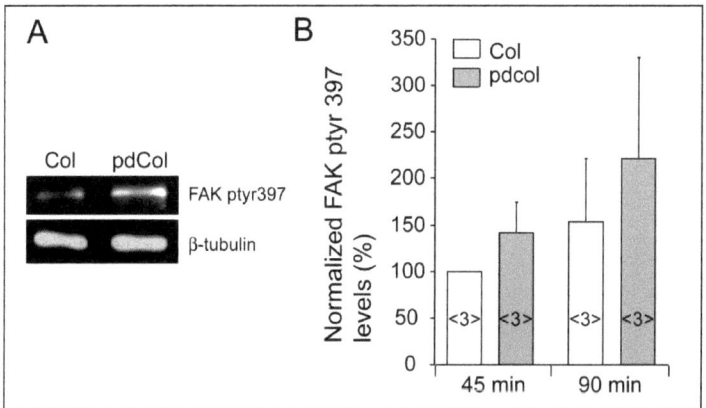

Fig. 56. FAK tyr397 phosphorylation on Col/pdCol. (A) Western blots showing FAK phosphorylation at tyr397. Cell lysates were taken 45 and 90 min after seeding MC3T3-E1 onto Col/pdCol. (B) FAK ptyr397 levels were quantified and normalized to loading controls (β-tubulin). Brackets indicate the number of western blots analysed.

FAK tyr397 phosphorylation was enhanced by 50 %/55 % in cells seeded on pdCol after 45 min/90 min (Fig. 56). Increased FAK phosphorylation might be attributed to differential integrin signalling. It has been shown that differential integrin binding, e.g. in cells plated on different adhesive ECM substrates (FN, collagen), differently stimulates FAK phosphorylation[389]. FAK phosphorylation generates a binding site for Src kinase and thereby triggers downstream signalling pathways, e.g. ERK and MAPK pathways[390, 391]. Beside other functions, FAK has been implicated in regulating cell migration and spreading[390, 391]. This might explain enhanced spreading and motility observed for pre-osteoblasts on pdCol. Recent studies suggested that FAK signalling regulates osteogenic differentiation of mesenchymal stem cells (MSC) via ERK and MAPK pathways[392]. Thus, it was hypothesized that Col and pdCol differently influenced osteogenic differentiation of MC3T3-E1 cells.

Proliferation and osteogenic differentiation on Col and pdCol

Cells were cultured under osteogenic differentiation conditions for 28, 35 and 42 d. Then matrix mineralisation, i.e. the extend of calcium phosphate deposition, was quantified (appendix B11). Mature osteoblasts secrete collagen type I and non-collagenous proteins, such as alkaline phosphatase, FN, osteocalcin, osteonectin, osteopontin and bone sialoprotein[393]. Calcium phosphate crystallization is induced by alkaline phosphatase under the control of osteonectin, osteopontin and osteocalcin[394, 395].

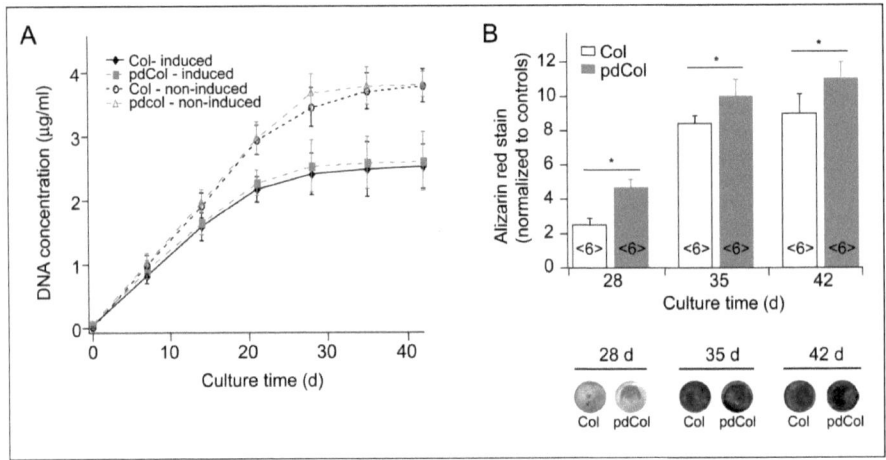

Fig. 57. Cell proliferation and matrix mineralisation on Col/pdCol. (A) Cell numbers were quantified after 7, 14, 21, 28, 35 and 42 d of cell culture. For each time point six samples were analysed. (B) Col/pdCol mineralisation was analysed by quantitative alizarin red S staining after 28, 35 and 42 d of cell culture. Staining intensity was normalized to staining of non-induced controls. Numbers in brackets indicate the number of experiments. Data are presented as mean±SD. Below, example photographies of stained 48 wells are shown (data was acquired by collaborators, Prof. D. Hutmacher).

Since matrix mineralisation occurs due to the action of mature osteoblasts, matrix mineralisation is a commonly used marker for osteogenic differentiation[396]. Matrix mineralisation was quantitatively assessed by staining cell layers with alizarin red S (appendix B11). During the first two weeks of culture, no significant amounts of deposited calcium were observed (not shown). This is explained by a lag phase during which the cells progress the process of becoming mature osteoblasts. Starting at four weeks of cell culture, strong matrix mineralisation was observed. Matrix mineralisation was significantly increased for pdCol compared to Col (Fig. 57 B). To exclude that increased mineralisation was an effect of potentially increased cell numbers, cell proliferation on Col and pdCol was compared. At all analysed time points no significant differences in cell numbers were found (Fig. 57 A). Comparing cell proliferation between cells that were cultured in presence and absence of osteoinductive medium it was observed that cells grown under osteogenic differentiation conditions ceased proliferation compared to uninduced cells. This further indicated that MC3T3-E1 cells underwent osteogenic differentiation.

Taken together, it was concluded that pdCol stimulated osteogenic differentiation of MC3T3-E1 cells. This finding was supported by further cell culture experiments performed on Col and pdCol coatings prepared on *thermanox* discs (see appendix A3). Whereas Col and pdCol exhibited on mica a fibrillar structure, collagen type I did not assemble into fibrils on *thermanox* discs (not shown). Hence it could be concluded that the effect of enhanced mineralisation was caused at the molecular level, not at the macromolecular level. In addition, our collaborators obtained comparable results in experiments using different cell types, e.g. primary human osteoblasts, and coatings of different collagen sources (not shown). This indicates that the effect of enhanced mineralisation found on pdCol is independent on cell type, collagen source and structural features of the coatings. These results are also in accordance with previous studies using human adult bone marrow stromal cells[378, 379] and human dermal fibroblasts[397]. In both studies an enhanced differentiation potential of cells cultured on denatured compared to native collagen type I was reported.

Several studies have demonstrated that integrin-ECM interactions can modulate osteogenic differentiation potential[370, 392, 398]. ERK and MAPK signalling pathways have been implicated in this process by activating multiple transcription factors and growth factors, for instance Runt-related transcriptional factors, Smads or bone morphogenetic protein- 2 (BMP-2)[392]. However, the exact mechanisms are not well understood. Since cells adhered via different integrin heterodimers to Col and pdCol, it was concluded that differential integrin-mediated signal transduction caused the differences in osteogenic differentiation. Recently a functional role of FN and its receptor $\alpha_5\beta_1$ in promoting osteogenic differentiation was demonstrated[398-402]. Since RGD-dependent integrins - including α_v and $\alpha_5\beta_1$- bound to pdCol, but not to Col, these integrins might have activated signal transduction pathways responsible for mineralisation initiation.

5.4 Conclusions & Outlook

In the presented work SCSF quantitatively showed that partially denaturing collagen type I leads to the exposure of cryptic RGD-motifs and enhanced adhesive interactions with pre-osteoblasts. α_v- and $\alpha_5\beta_1$- integrins were identified as predominant integrins recognizing pdCol. RGD-dependent binding mechanisms of cells to denatured collagen I have been already demonstrated in previous studies[358, 364, 365]. However, most previous studies applied adhesion assays that are limited in a way that they do not quantitatively measure cell adhesion forces.

Concomitantly with increased adhesion, pdCol promoted spreading and migration of pre-osteoblasts. This was in accordance with studies showing that migration of human dermal fibroblasts cells was enhanced on denatured compared to native collagen[397]. Enhanced cell attachment and migration on denatured collagen may present a mechanism by which cells are guided to sites of collagen degradation. Such guidance may assist removal of degraded collagen after tissue injury or during tissue remodelling to accomplish tissue homeostasis. Indeed, matrix remodelling was previously reported to be increased when fibroblasts were grown on denatured instead of native collagen I[373].

In the presented work denaturation of collagen type I was caused by thermal denaturation. Can such thermal denaturation be correlated with collagen degradation occurring *in vivo*? It has been demonstrated that proteolytic degradation of collagen type I destabilizes its triple-helical structure leading to unwinding of the triple helices at physiological temperatures[362]. Thus, both thermal denaturation and proteolytic degradation of collagen may induce similar local unfolding of the collagen molecules. This suggests that the mechanisms by which cells react to thermally denatured and proteolytically degraded collagen *in vivo* may be similar. It might be interesting to analyse in a future study the effect of different collagenases on Col matrices and on cell attachment and behaviour.

Pre-osteoblasts cultured on pdCol exhibited enhanced differentiation kinetics and potential. Since integrin signalling was altered upon collagen denaturation, molecular mechanisms of cell adhesion to native and partially denatured collagen I may directly correlate with enhanced osteogenic differentiation. It appears surprising that initial matrix contact can trigger such long-lasting effects over six weeks of cell culture. This can be explained by a kind of feedback loop implemented in the process of cell-ECM interactions. There exists a bi-directional interplay between integrins and the ECM [403-405] (Fig. 58). As discussed in 1.2.3, specific ligation to ECM ligands triggers intracellular signalling pathways. Besides receiving and transmitting chemical signals, integrins are also mechanical linkers between contractile elements in the cell interior (acto-myosin cytoskeleton) and the ECM. Integrin association with mechano-sensitive proteins enables cells to sense the mechanical properties of their surrounding and to react accordingly[406-408]. Thus, together with altered chemical signals, different mechanical properties of the surrounding matrix (as seen for Col and pdCol) might affect cellular behaviour. In response, cells can modify composition and mechanical properties of their ECM by adding new ECM proteins or by secreting proteases degrading the ECM. Different signals initially received by the cells grown on Col and pdCol may have lead to alternative ECM modifications. For instance, depending on the matrix type (Col, pdCol), distinct ECM proteins might have been secreted by the cells. Since this newly deposited matrix stimulates cellular behaviour inversely, long-lasting signals can thereby be produced. Ongoing studies might unravel differences in ECM composition after prolonged culture periods.

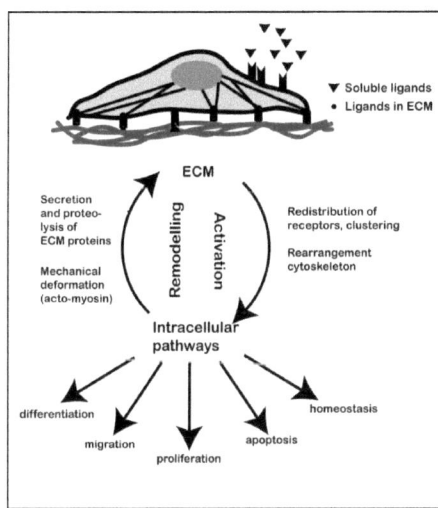

Fig. 58. Reciprocal interactions occurring between cell and ECM. Cell receive signals from soluble and matrix-bound molecules that bind to cell-surface receptors. This activates intracellular signalling pathways. Vice versa the cell can act back on the ECM modifying it by secretion/ proteolysis of ECM proteins or by applying mechanical stress on the matrix hence deform it. The modified ECM sends back altered signals to the same cell and neighbouring cells. The figure was adapted with slight changes from [405].

An important issue in regenerative medicine is the development of optimised systems for the expansion of stem cell *ex vivo*. A common problem is the gradual decrease of cellular differentiation potential. Thus *ex vivo* culture conditions have to be optimised to preserve cellular functions[373]. The presented findings may present a basis to exploit the use of denatured collagen in various cell-expansion scenarios for tissue engineering applications. Thereby the loss of proliferative capacity and differentiation potential of the cells in use might be prevented.

In this work well-characterized collagen matrices consisting of parallel, highly ordered collagen fibrils were used. These collagen fibrils exhibited the characteristic D-band, thus shared similarities to *in vivo* assembled fibrils. In numerous studies cell adhesion to non-fibrillar collagen coatings was studied. However, it was shown that cell adhere to monomeric and fibrillar collagen type I via different integrins[409]. Analysing cellular interactions with fibrillar collagen has the advantage of being more comparable to collagen assembled in *vivo*. Moreover, the used Col matrices present highly ordered pattern of integrin binding sites. Since it has been shown that ligand spacing has an important influence on adhesion, spreading, proliferation and differentiation of cells[410, 411] the used matrices may promote cell attachment and growth. However, it was observed that MC3T3-E1 cells disassembled Col and pdCol within few hours and the homogenous structure of the matrix is thereby lost. In contrast to native collagen type I fibrils, collagen fibrils within Col matrices are not cross-linked (1.1.1); such crosslinks significantly improve the mechanical stability of the matrix. Thus, to preserve the structure of highly ordered collagen coatings over longer cell culture periods, chemical cross-linkers or enzymatic cross-linking might be tested in future work to improve long-term stability of Col matrices.

Chapter 6

Quantifying adhesion of myeloid progenitors to bone marrow derived stromal cells

6.1 Abstract

Expression of the fusion protein BCR/ABL is hallmark of chronic myeloid leukemia (CML). BCR/ABL is a constitutively active tyrosine kinase interfering with normal cell proliferation, apoptosis and differentiation. To which extend and by which mechanisms BCR/ABL influences the adhesion of leukemic cells to bone marrow stromal cells (BMSC) is not clear. In the study SCFS was applied to quantify adhesion of BCR/ABL transformed 32D cells (32D-BCR/ABL) to the BMSC line M2-10B4. SCFS data revealed that adhesion forces of 32D-BCR/ABL cells were three fold increased compared to control 32D cells (32D-V). The BCR/ABL-mediated effect of enhancing cell adhesion could be reversed to control levels by imatinib mesylate (IM), an inhibitor of the BCR/ABL tyrosine kinase activity. SCFS further showed that adhesion forces to FN and collagen type I were higher for 32D-BCR/ABL cells than controls, suggesting that β_1-integrin plays a major role in mediating adhesion of leukemic cells to BMSC. Indeed, a β_1-integrin blocking antibody nearly abolished adhesion of 32D-V and -BCR/ABL cells to BMSC. Flow cytometry and western blot analysis revealed significantly increased β_1-integrin concentrations on the surface of 32D-BCR/ABL cells. Since no significant differences of β_1-integrin mRNA levels were detected, it was suggested that β_1-comprising integrin heterodimers were regulated post-transcriptionally by BCR/ABL. The presented data indicate that specifically interfering with β_1-integrins might optimize CML therapies by circumventing effects of cell adhesion mediated drug resistance.

6.2   Introduction

Leukemias are cancers of the blood and bone marrow that are characterized by abnormal proliferation of white blood cells[412]. Critical for the development of leukemias are genetic alterations occurring in pluripotent hematopoietic stem cells. Four major types of leukemias have been defined, namely chronic myeloid leukemia (CML), acute myeloid leukemia (AML), chronic lymphoid leukemia (CLL) and acute lymphoid leukemia (ALL). Their classification is based on the (myeloid or lymphoid) origin of the mutated precursor cell and on the course of the disease. The current project focuses on CML, a clonal disorder of myeloid progenitor cells originating in the bone marrow. With 1-2 new cases per 100 000 cases per year, CML is considered a relatively rare disease, the median age of diagnosis is 53 years[413]. The course of CML can be divided into three phases, an initial chronic phase characterized by an elevated white blood cell count[413] that can persist for several years. The disease either passes through an accelerated phase with immature blood cells accumulating in the peripheral blood or directly progress to blast crisis, the terminal stage. Blast crisis resembles acute myeloid leukemia and is characterized by an excess of immature malignant precursors in the blood. Without treatment the median survival time of patients after diagnosis is about 4 to 6 years[414].

In most types of cancers and also other types of leukemias several cytogenetic alterations have to account for development of the disease. In CML, however, formation of the so-called Philadelphia chromosome (Ph chromosome)[415] is considered to be the critical event leading to oncogenesis. The Ph chromosome was firstly described in 1960 by Peter Nowell and David Hungerford as chromosomal abnormality occurring in 95%[1] of CML patients [415]. The reasons for the formation of the Ph chromosome are not clear, the molecular pathogenesis of CML, however, is well understood[417]. The Ph chromosome is consequence of a reciprocal translocation between chromosomes 9 and 22, t(9;22)(q34;q11)[418, 419]. This translocation leads to juxtaposition of sequences of the proto-oncogene Ableson leukemia virus (encoding for c-Abl) on chromosome 9 and the breakpoint cluster region (Bcr) on chromosome 22[420]. As a result the fusion genes BCR/ABL and ABL/BCR are formed[421]. There are not many reports on biological relevance of the ABL/BCR, differently to the product of BCR/ABL that plays a central role in cell transformation. Depending on the exact fusion site of exons, BCR/ABL transcripts are translated into three major isoforms, a 210kDa protein ($p210^{BCR/ABL}$)[422], the most common type, a 190kDa protein ($p190^{BCR/ABL}$) or more rarely 230kDa protein ($p230^{BCR/ABL}$)[413]. It was demonstrated that

---

[1] Whether the 5% of cases in which no Ph chromosome is detectable (usually referred to as atypical or Ph negative CML) have to be classified separately is controversial[416].

p210$^{BCR/ABL}$ can transform a variety of hematopoietic cell types *in vitro*[423] and *in vivo* in animal models[424-427].

c-Abl is a tightly regulated non-receptor tyrosine kinase that plays a role in actin reorganization, cell proliferation and survival/apoptosis[428]. c-Abl is located at the nucleus and in cytoplasm of the cell. Upon DNA damage c-Abl is activated, translocated into the nucleus and induces apoptosis. In contrast, cytoplasmic c-Abl is linked to growth factor receptor signalling pathways and promotes survival and proliferation of the cells. In that way the location of c-Abl decides whether such opposite events as survival or apoptosis occur. The oncogenic effects produced by mutated Abl are mediated by its exclusive cytoplasmic localization together with its increased tyrosine kinase activity. The tyrosine kinase activity of c-Abl is normally tightly regulated by interactions with negative regulators, such as F-actin[429], phosphoinositides [430, 431] and by intramolecular autoinhibition mechanisms. In contrast, p190$^{BCR/ABL}$ and p210$^{BCR/ABL}$ are auto-phosphorylated which is critical for cellular transformation[432]. In normal hematopoiesis cytokines and other extracellular stimuli regulate proliferation, apoptosis and differentiation of hematopoietic progenitor cells via intracellular signalling pathways. These signalling cascades include Ras/Raf-1/Erk and PI3K/Akt as well as JAK/STAT5 (JAK: Janus family of tyrosine kinases, STAT: signal transducers and activators of transcription)[432-435]. IL-3, a soluble highly glycosylated 26kDa protein, is such a cytokine that promotes cell growth and division and inhibits pathways leading to apoptotic cell death[435].

In BCR/ABL transformed cells, these pathways are constitutively activated by BCR/ABL and consequently cells become growth factor independent. It has been shown that BCR/ABL activates Ras/Raf-1/Erk and PI3K/Akt signalling pathways as well as JAK/STAT5[434-438](Fig. 59). Deregulated activation of these pathways promotes increased cell proliferation, decreased apoptosis and growth factor independence, respectively[435]. Decreased cell death and increased proliferation lead to a massive clonal expansion of progenitor cells occurs during CML[435, 436, 437, 439].

There are also other malignancies with deregulated Abl activity. For example, v-Abl, a virally encoded form of Abl, causes preB-cell leukemias in mice[428]. Recent data suggest that increased activation of c-Abl tyrosine kinase can also be involved in malignant solid tumors of lung and breast[428, 440].

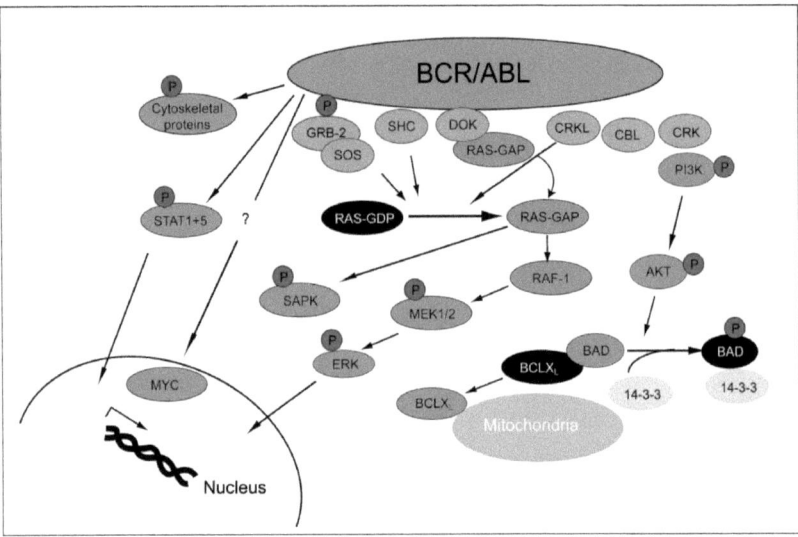

Fig. 59. Signalling pathways affected by BCR/ABL. Note that there are more interactions between BCR/ABL and signalling proteins that have been neglected for simplicity. Taken with slight modification from[441].

Historically, radiation was applied as first treatment against CML. Later on, so called anti-metabolites were used, these include cytoarabine, hydroxyurea, alkylating agents, interferon alpha and steroids have been applied in the treatment of CML. Stem cell transplantation represents the only method by that CML patients can be completely cured, however, its application is often limited by the advanced age of patients[442]. The latest developments include targeted therapies that apply small molecules that specifically interfere with target molecules responsible for oncogenesis. Targeted therapies have the advantage over anti-metabolites that they can be more effective and also have less side effects. Examples for targeted therapy are for instance tyrosine kinase inhibitors and monoclonal antibodies. In the nineties a large random screen for several tyrosine kinases was performed. The 2-phenylaminopyrimidines were first reported as potent inhibitors of tyrosine kinases with high selectivity for Abl and PDGF-R tyrosine kinases[443]. To optimize 2-phenylaminopyrimidines for optimized inhibition of PDGF-R, chemically related compounds were synthesized. The most potent molecules in the screen were all inhibitors of both, v-Abl and the

PDGF-R kinases. *STI 571* or *imatinib mesylate* (IM) originates from these trials and was the lead compound for preclinical development[443, 444]. IM is historically of particular importance, as it was the first drug developed for targeted therapies. In the last years IM has become medicament of choice for the treatment of CML. IM effectively inhibits the BCR/ABL tyrosine kinase activity by binding competitively at its ATP-binding site (Fig. 60). Adverse events with IM are usually low[443].

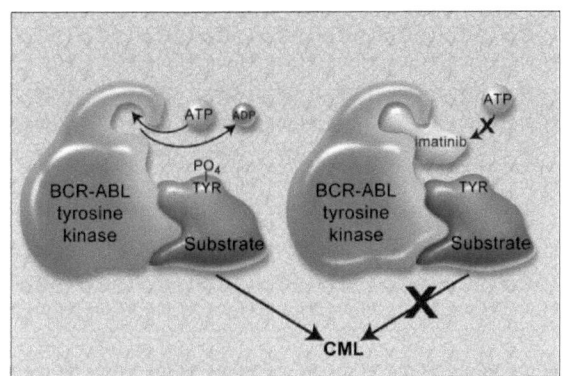

Fig. 60. Mechanism of IM. The constitutively ac-tive tyrosine kinase BCR/ABL transfers phos-phate from ATP hydrolysis to tyrosine residues of different substrates activa-ting signalling pathways critical for CML (left). IM blocks ATP binding to BCR/ABL and thereby inhibits its tyrosine kinase activity (right). Taken from[445].

In some cases problems of disease persistence and resistance arise. Therefore alternate compounds have been developed, such as nilotinib and desatinib (a Src inhibitor) that are tested in clinical trials[446, 447]. Alternatively other types of inhibitors might be used that act downstream of BCR/ABL, for instance inhibitors of Ras/Raf/MEK/ERK or PI3K/PTEN/Akt/mTOR pathways[448, 449] (Fig. 59). These pathways have been targets in cancer therapies since they play a critical role in promoting proliferation. Respective inhibitors have been developed and applied for different types of cancers. These could be useful for IM-resistant CMLs, clinical trials are planned [449].

Normal hematopoiesis is guided by dynamic interactions between hematopoietic stem cells and the bone marrow microenvironment. The bone marrow microenvironment is composed of stromal cells and extracellular matrix components, such as FN[450, 451] and collagens. Bone marrow cells secrete soluble factors controlling proliferation and differentiation of hematopoietic cells. However, beside soluble factors, direct interactions between hematopoietic stem cells and bone marrow cells via adhesion molecules occur. Those interactions have been also suggested to be critical to a phenomenon referred to as cell-adhesion mediated drug resistance (CAM-DR). Thereby integrins on leukemic cells bind to bone marrow cells or to secreted ECM proteins and activate anti-apoptotic signalling pathways. Recently, a crucial role in CAM-DR has been proposed for β_1-integrin[452] and, more specifically, for interactions between $\alpha_4\beta_1$-integrins on leukemic cells and VCAM-1 on bone marrow cells or FN. Several studies have also suggested an anti-apoptotic effect of the bone marrow microenvironment *in vivo*: Matsunaga et al. demonstrated in a murine model for AML that bone marrow cells protect leukemic cells from the chemotherapeutic drug cytoarabine, by inhibiting apoptosis via Bcl-2 and Bcl-X[453]. Furthermore combinations of blocking antibodies directed against integrins (VLA-4) and chemotherapeutic drugs could increase survival of diseased mice. Thus, integrin signalling represents an interesting therapeutic target when combined with chemotherapeutic drugs. This might involve antibodies interfering with integrin function or inhibitors acting more downstream.

The named examples point out, that it is of great importance to understand the molecular basis underlying the interactions between myeloid cells and BMSC. A better knowledge about the involved molecules may aid the development of targeted therapies. Several published studies have addressed the influence of BCR/ABL on cell adhesion. Commonly cell adhesion to FN model surfaces was analysed, only a few studies have addressed cell adhesion to BMSC[454, 455]. The performed studies gave controversial results[456-458]. While adhesion of BCR/ABL-transformed hematopoietic cell lines, including 32D cells, to FN coated surfaces was found to be increased in some studies[456, 457, 459], others observed reduced adhesion[460-462]. The inconsistent results are presumably attributable to the properties of the modified cells and the experimental strategy being used. None of these studies applied cell adhesion assays that characterized cell-cell adhesion quantitatively. Thus, objective methods to quantitatively characterize adhesion of leukemic cells are clearly needed.

In this work the effect of BCR/ABL on adhesion of myeloid progenitor cells to a BMSC cell line (M2-10B4) or ECM proteins (FN, Col) was quantitatively investigated. For that purpose, adhesion of 32D cells that were retrovirally transfected to stably express BCR/ABL (32D-BCR/ABL cells) was quantitatively compared to control cells transfected with empty vector (32D-V cells).

6.3 Results & discussion

6.3.1 Quantifying cell adhesion between 32D and BMSC

Our collaborators had performed washing assays to compare the attachment of adherent 32D-V and 32D–BCR/ABL cells to BMSC. A significantly higher percentage of 32D-BCR/ABL cells adhered to BMSC than of 32D-V control cells (t-test: $p < 0.001$) (Fig. 61 A, B). Importantly, depending on incubation time and concentration of IM the increased percentage of adherent 32D-BCR/ABL cells could be reduced to that of 32D-V cells. In contrast IM showed no effect on the percentage of 32D-V cells attached to BMSC.

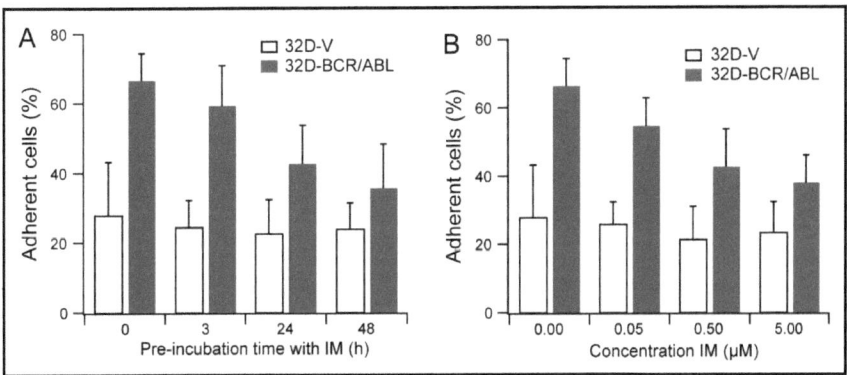

Fig. 61. Quantification of cell-cell adhesion using washing assays. 32D-V and 32D cells (-V and - –BCR/ABL cells) were pre-incubated with or without IM prior to their co-culture with BMSC. (A) 32D cell adhesion to BMSC depending on the IM pre-incubation time. (B) 32D cell adhesion to BMSC depending on IM concentration.

Since washing assays cannot provide quantitative data on cell adhesion forces adhesion of 32D cells to BMSC was subsequently analysed by SCFS. Individual 32D cells were attached to the AFM cantilever and F-D curves were recorded on BMSC grown on glass coverslips (Fig. 62). For all analysed contact times (5, 30, 120 sec) detachment forces of 32D-BCR/ABL cells lied significantly ($p < 0.001$) above that of 32D-V cells. At a contact time of 120 sec 32D-BCR/ABL cells showed a median detachment force of 3317 ± 2169 pN, an approximately three-fold higher detachment force than that observed for 32D-V cells (998 ± 362 pN).

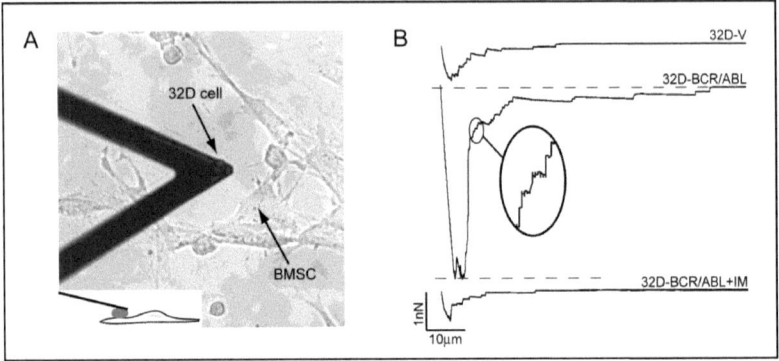

Fig. 62. Experimental setup to quantify cell-cell adhesion by SCFS. (A) Phase contrast image of a 32D cell attached to a tippless AFM cantilever. Below, a sparse layer of BMSC is visible. The small inset (bottom left) illustrates the sideview during the contact between a single 32D cell and a stromal cell during a F-D cycle. (B) Representative F-D curves recorded for 32D-V control cells, 32D-BCR/ABL cells and for 32D–BCR/ABL cells incubated with 0.5mM IM for 18-22h. (contact time 120sec).

Pre-incubating 32D-BCR/ABL cells with 0.5 µM IM for 18 – 22 h reduced detachment forces to similar values (889 ± 505 pN, 120 sec) as observed for untreated 32D-V cells (Fig. 63 A). Detachment forces of 32D-BCR/ABL cells (Fig. 63 B) showed a wider distribution indicating that their adhesion varied more compared to that of the 32D-V control cells.

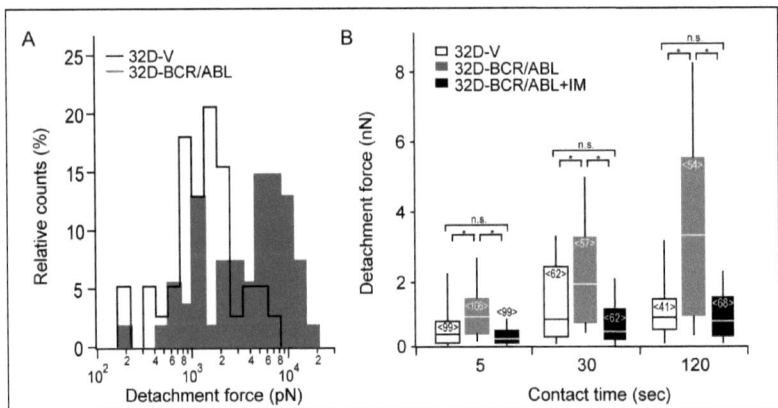

Fig. 63. Quantification of cell-cell adhesion by SCFS. (A) Detachment forces for 32D-V and 32D-BCR/ABL after a contact time of 120sec. (B) Detachment forces for 32D cells at contact times of 5, 30 or 120 sec. Data are presented as boxplots. Inserted numbers within bars (<n>) indicate the total number of analysed force curves. Brackets above the bars represent the results of a statistical test (Mann- Whitney) (* p < 0.05, n.s. ≥ 0.05).

Both, washing assays and SCFS, revealed that 32D cells expressing BCR/ABL fusion proteins adhered significantly stronger to BMSC compared to 32D-V control cells. Pre-incubation of 32D cells with IM reversed the increased adhesion of 32D-BCR/ABL cells to BMSC to that of 32D-V cells. Since 32D-V and 32D-BCR/ABL cells differed only in the expression of the BCR/ABL oncogene and IM could reverse enhanced adhesion of 32D-BCR/ABL, it was concluded that the enhanced cell-cell adhesion was consequence of the tyrosine kinase activity of BCR/ABL. To analyse the adhesive interactions in more detail, in the next step single rupture events in F-D curves were analysed.

6.3.2 Analyzing single rupture events in F-D curves

Next the magnitude of single rupture events ("j") (Fig. 62 B, Table 4) was analysed in F-D curves recorded with 32D-V and 32D-BCR/ABL on BMSC after 5 sec contact. The most frequent force was 53 ± 35 pN for 32D-BCR/ABL and similar for 32D-V cells (47 ± 32 pN). These results suggested that the binding strength of most unbinding events remained the same for both cell types. Incubation of 32D-BCR/ABL cells for $18-22$ h with 0.5 µM IM, decreased the most frequent rupture force to 37 ± 24 pN. A similar effect was observed in 32D-V cells in presence of IM. IM reduced single rupture events to similar forces (36 ± 24 pN). Since the effect of reduced single rupture events was also seen in 32D-V, this indicated that IM had further effects on myeloid progenitor cells than blocking BCR/ABL tyrosine kinase activity. It is known that IM inhibits - besides BCR/ABL- c-Abl, c-kit and platelet derive growth factor receptor (PDGF). However, it may be speculated that interfering with c-Abl that locates to adhesion sites and is involved in actin cytoskeleton organization[429], alters the mechanical properties of the adhesion receptor-cytoskeleton linkage.

F-D curves of 32D-BCR/ABL cells exhibited about 47 % more force jumps than F-D curves recorded with 32D-V cells (Table 4). Addition of 0.5 µM IM to 32D-BCR/ABL cells reduced the number of single rupture events from 7.8 ± 6.0 to 5.5 ± 3.8, a similar value as found for 32-V control cells (5.3 ± 5.0). IM had no effect on the number of single unbinding events observed for 32D-V cells (not shown). Thus, it was concluded that 32D-BCR/ABL cells exhibited more active adhesion molecules compared to 32D-V cells.

Adhesion to	Cell type / condition	<Number of jumps> ± SD	Most frequent jump size [pN] ± width	N	n
BMSC	32D-V	5.3±5.0	47±32	79	421
	32D-BCR/ABL	7.8±6.0	53±35	64	498
	32D-BCR/ABL+IM	5.5±3.8	37±24	77	422
Fibronectin	32D-V	4.2±2.6	72±28	42	178
	32D-BCR/ABL	9.1±3.9	67±36	43	391
	32D-BCR/ABL+IM	4.2±2.5	72±44	23	97
Collagen I	32D-V	2.6±1.4	51±30	27	71
	32D-BCR/ABL	5.8±2.4	52±29	30	175
	32D-BCR/ABL+IM	3.8±1.8	48±22	34	128

Table 4. Analysis of single force jumps for cell-cell and cell-ECM measurements. Mean numbers of single force jumps (±SD) per F-D curve for 32D-V cells, 32D-BCR/ABL cells, and 32D-BCR/ABL cells after incubation of 0.5 µM IM for 18 – 22 h (contact time 5 sec).

6.3.3 Analyzing 32D cell adhesion to FN and collagen type I coated surfaces

Next, adhesion of 32D-V and 32D-BCR/ABL cells to FN-coated surfaces and collagen type I matrices (Col) was quantified. Compared to 32D-V cells, 32D-BCR/ABL cells showed significantly higher detachment forces (Fig. 64 A, B) on both FN and Col. Concomitant with the elevated overall adhesion to FN, for 32D-BCR/ABL cells more unbinding events (9.1 ± 3.9, 5 sec) were detected than for 32D-V cells (4.2 ± 2.6, 5 sec) (Table 4). The magnitude of single unbinding events was similar for 32D-BCR/ABL (67 ± 36 pN) and 32D-V control cells (72 ± 28pN). This suggested that 32D-BCR/ABL cells exposed an increased number of active adhesion molecules. Concomitant with this observation, 32D-BCR/ABL cells increased their detachment forces with contact time while 32D-V cells did not. Pre-incubation of 32D-BCR/ABL cells with 0.5 µM IM for 18 to 22 h significantly reduced detachment forces (Fig. 64 A, B) and the number of rupture events j (Table 4). In general, detachemnt forces of 32D cells were found to be much higher on FN than on Col, which coincided with an elevated number of single rupture events.

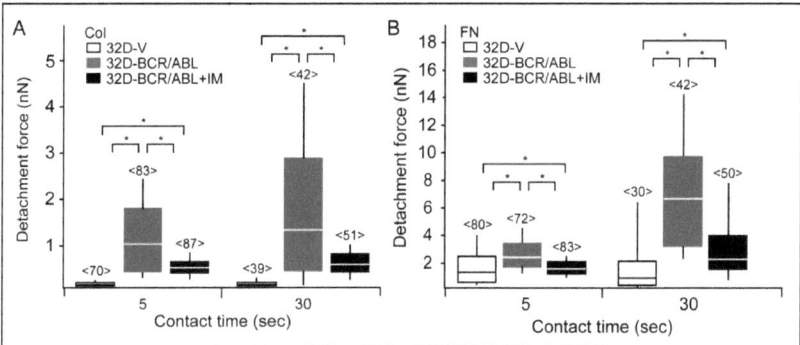

*Fig. 64. **Quantifying adhesion of 32D cells to ECM proteins.** Adhesion of 32D (-V and – BCR/ABL) to Col **(A)** or FN **(B)**) coated surfaces was quantified for contact times of 5 or 30sec. Data are presented as boxplots. The results of a Mann-Whitney test are shown.*

Above results indicate that adhesion of 32D-BCR/ABL to BMSC, but also to FN and Col was increased. As mentioned above, there are conflicting reports on the effects of BCR/ABL on cell adhesion. Whereas several research groups found enhanced adhesion of BCR/ABL positive cells to FN [456, 457, 459], others reported a loss of adhesion [460-462]. These controversial results might be explained by different culture conditions and the respective cell model used[463]. Indeed, most of the studies that used human BCR/ABL positive CD34$^+$ progenitor cells showed decreased adhesion to FN, whereas other studies using the murine cell lines 32D or BAF3 reported an increased adhesion to FN upon BCR/ABL expression. However, there is no clear indication that cell adhesion varies in a species-specific manner. Hence, additional factors as the expression level of BCR/ABL may define the effect on cell adhesion. Barnes et al. demonstrated that low levels of BCR/ABL decreased adhesion to FN, cells whereas high levels lead to enhanced adhesion[464]. The cellular system used in this study compares favorably with the high expression clones reported by Barnes et al.[464] since BCR/ABL expression and tyrosine kinase activity were strong (immunoprecipitation assays of BCR/ABL phosphorylation, data not shown). Typically such increased levels of BCR/ABL expression have been observed in BCR/ABL positive blast crisis of CML patients[459, 465].

None of the previous studies could correlate enhanced adhesion of BCR/ABL expressing cells with its tyrosine kinase activity[457, 466]. Even though these experiments applied much higher concentrations of IM (10μM) for a time period of 15 – 17 h, no effect on the adhesion of BCR/ABL expressing cells could be observed. In contrast to these experiments, SCFS measurements showed that incubation with 0.5 μM IM over a time period of 18 – 24 h was effective to reduce adhesion of 32D-BCR/ABL cells to that observed for 32D-V control cells. The concentration of 0.5 μM IM lies close to that found in the blood of CML patients (≤1 μM) receiving IM as a therapeutic drug[467, 468].

It might be assumed that the strongest IM mediated inhibition of 32D-BCR/ABL cell adhesion was reached after a pre-incubation time of 24h. Thus, a time period of 15 – 17 h may not be sufficient to observe an effect of IM on the BCR/ABL induced cell adhesion. The obtained controversial results might be further attributed to the non-quantitative adhesion assay used.

Increased adhesion of BCR/ABL expressing 32D cells FN and Col suggested that receptors of the integrin family were responsible for the amplified adhesion of 32D-BCR/ABL cells. As β_1-integrins are involved in cell binding to FN (mainly $\alpha_4\beta_1$- and $\alpha_5\beta_1$-integrin) and Col (predominantly through $\alpha_2\beta_1$-integrin), it became the primary target in the following experiments.

6.3.4 Role of β_1-integrins in mediating adhesion of 32D cells to BMSC

To analyse the role of β_1-integrin, 32D-V and 32D-BCR/ABL cells were pre-incubated for 1 h with 20 µg/ml of the β_1-integrin blocking antibody Ha2/5[469]. Characterizing these Ha2/5-treated cells by SCFS revealed that adhesion of 32D-V cells and of 32D-BCR/ABL cells to BMSC was almost completely abrogated (Fig. 65). For 32D-BCR/ABL cells, the median detachment forces went down from 1016 ± 500 pN to 269 ± 55 pN after a contact time of 5 sec.

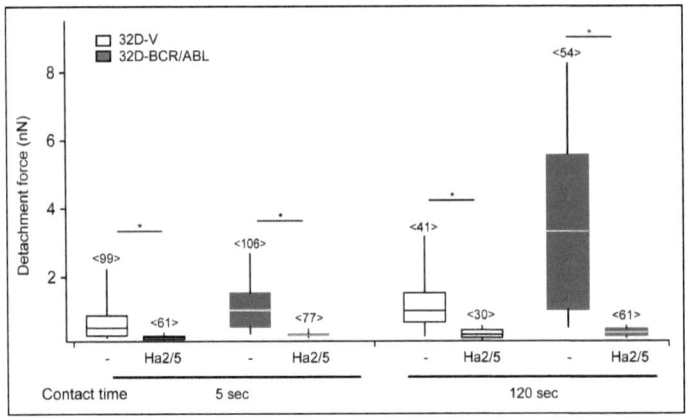

Fig. 65. Quantifying adhesion of 32D cells to BMSC blocking β1-integrins. To inhibit β₁-integrin function, 32D-V and 32D–BCR/ABL cells were pre-incubated for 1 h with 20µg/ml blocking antibody Ha2/5. Subsequently adhesion to stromal cells was measured for contact times of 5 or 120 sec. Data are presented as boxplots. The number of analysed F-D curves is indicated in brackets. Results of a Mann-Whitney test are shown.

In presence of Ha2/5, the detachment forces of 32D-V cells after 5 sec of contact decreased from 512 ± 272 pN to similarly low values of 235 ± 49pN. This indicated that β_1-integrin significantly contributed to the adhesion of 32D cells to BMSC. Since detachment forces of 32D-BCR/ABL and 32D-V cells were diminished to similar forces, β_1-integrin was concluded to be responsible for the BCR/ABL enhanced adhesion. The fact that cell adhesion was not completely abrogated by blocking β_1-integrin might indicate that further adhesion molecules were involved at a low level. Alternatively, it is possible that β_1-integrins were not completely blocked by the antibody. These results could be confirmed qualitatively, performing washing assays after pre-incubation of 32D-V and 32D-BCR/ABL cells for 1 h with 20 µg/ml of the β_1-integrin blocking antibody Ha2/5 and 3 h of co-culture (not shown). This indicated that most interactions of 32D cells and BMSC involved β_1-integrin both during early adhesion events within 2 min, as monitored by SCFS, but also in prolonged adhesion processes as observed by washing assays.

SCFS further showed that the absence of Mg^{2+} and/or Ca^{2+} ions in the medium significantly reduced adhesion of 32D-V and 32D-BCR/ABL cells to BMSC (not shown). This was in agreement with above findings, since several integrin heterodimers require both Mg^{2+} and Ca^{2+} ions for ligand binding[470, 471].

SCFS is a functional method that detects unbinding of adhesion molecules that are in an active ligand-binding conformation. Since SCFS showed that cell-cell adhesion was dominated by integrins comprising subunit β_1, this pointed to an increased number of active β_1-integrin molecules present on 32D-BCR/ABL cells. However, SCFS does not permit to distinguish if a higher total number of β_1-integrins was present, or if a higher proportion of β_1-integrins was in a high-affinity conformation. This had to be tested by complementary experiments, such as flow cytometry and western blots.

6.3.5 β_1-integrin protein and mRNA levels

Analyzing protein concentrations by flow cytometry revealed that 32D-BCR/ABL exposed higher levels of cell surface associated β_1-integrin compared to 32D-V. This increased level of β_1-integrin expression could be reversed upon addition of IM in a time (not shown) and concentration dependent manner (Fig. 66 B, C).

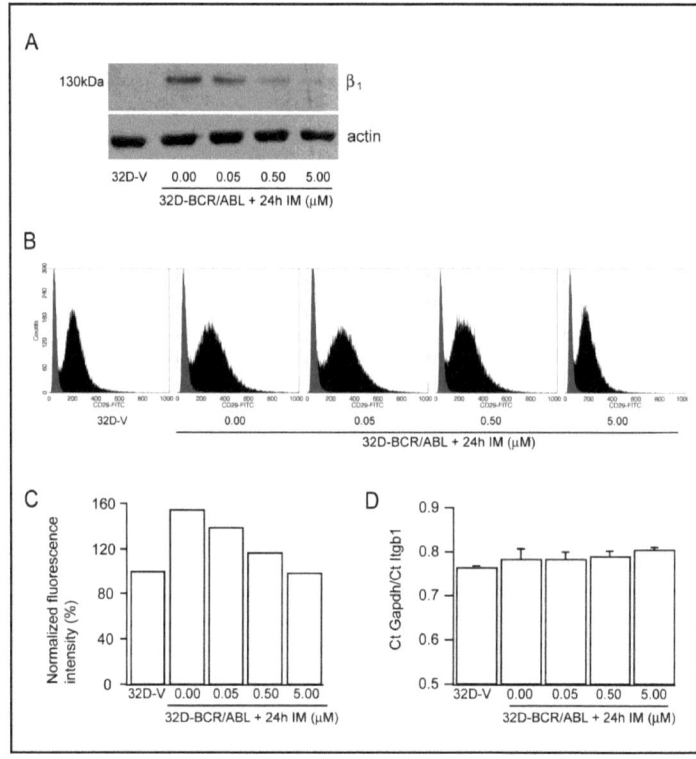

Fig. 66. β₁-integrin and gene expression in 32D cells. **(A)** *Itgb1 (β₁-integrin gene) mRNA transcripts were evaluated by quantitative real time-PCR. Glyceraldehyde-3-phosphate dehydrogenase (Gapdh) was used as house keeping.cell surface b1-integrin (CD29) was characterized by flow cytometry (black). Isotype controls are shown in grey. Mean fluorescence intensities are shown in the upper right corners.* **(B)** *Concentrations of mature β₁-integrins analysed by western blots. Actin was used as loading control.*

Similarly western blot analysis showed increased concentrations of mature 130 kDa $β_1$-integrin on the cell surface (Fig. 66 A). When carbohydrate moieties were digested by PNGase F, similar $β_1$-integrin concentrations were found for 32D-V and 32D-BCR/ABL indicating that cell-surface associated $β_1$-integrin, but not total concentrations were changed (not shown). In agreement with this observation, no differences in expression of the $β_1$-integrin encoding gene (Itgb1) between 32D-V and 32D-BCR/ABL cells were detected by real time PCR (Fig. 66 D).

Differing β_1-integrin concentrations at the cell surface might be due to differences in concentrations of α-integrin combining with β_1-integrin. β_1-integrins are usually synthesized in excess, whereas α-integrin subunits are rate limiting for the formation of different integrin heterodimers[472]. Twelve different α-integrins were described to form heterodimers with β_1-integrins[473]. Among these, expression of genes encoding for α_1-, α_2-, α_3-, α_4-, α_5-, α_6-, α_6-, α_7- and α_v-integrins was found in 32D (-V and –BCR/ABL) cells. By flow cytometry surface concentrations of α_4-, α_5- and α_v-integrins were analysed in 32D (-V, -BCR/ABL and -BCR/ABL + IM) cells. These three α-integrins are major fibronectin receptors. Compared to 32D-V cells 32D-BCR/ABL cells showed a lower α_4-integrin concentration and an increased α_v-integrin concentration. However, concentration differences could not be reversed by IM. Thus, different levels of α_4- and α_v- integrins cannot explain the observed differences in cell adhesion. In contrast, enhanced concentrations of α_5-integrin were detected at the surface of 32D-BCR/ABL cells. Similarly to β_1-integrin (Fig. 67), the increase of the α_5-integrin concentrations could be partially reversed upon pre-incubation with 0.5 µM IM for 18 h (Fig. 67). α_5-integrin exclusively combines with β_1-integrin to form $\alpha_5\beta_1$. Since $\alpha_5\beta_1$-integrin is a FN receptor, this might explain enhanced adhesion of 32D-BCR/ABL to FN, but not increased adhesion to Col, because Col and FN binding integrins are mutually different. Thus, it might be suggested that concentrations of a certain set of integrin heterodimers were enhanced in BCR/ABL expressing 32D cells. Previous studies have also found an increase of α_5- and β_1-integrin concentrations in BCR/ABL expressing 32D cells[456] which is in accordance with the presented data. Flow cytometry revealed lower α_4-integrin and consequently $\alpha_4\beta_1$-integirn concentration on the cell surface of 32D-BCR/ABL cells.

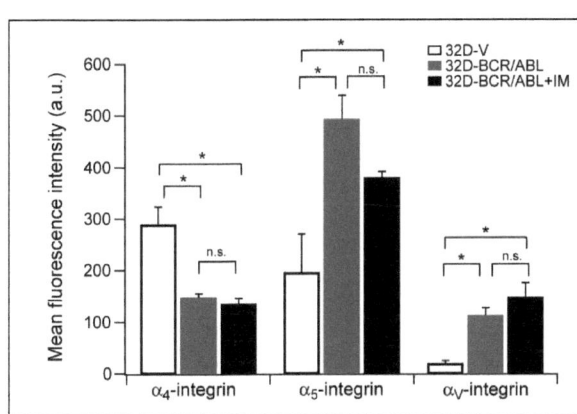

Fig. 67. Effect of BCR/ABL on 32D cell surface concentrations of integrins. Mean fluorescence intensities (+/- SD) (a.u.=arbitrary unit) of α_4-, α_5- and α_v-integrin-labelled 32D cells. Brackets above the bars show the results of a one-way ANOVA t-test test (* p<0.05, n.s. ≥ 0.05).

Considering that $\alpha_4\beta_1$-integrins are the only integrins mediating cell-cell adhesion ($\alpha_4\beta_1$-VCAM-1), it appears surprising that adhesion of 32D-BCR/ABL cells to BMSC was enhanced. Thus, it was hypothesized that interactions between 32D cells and BMCS were not primarily due to integrin mediated cell-cell interactions (VLA-4-VCAM-1), but dominated by other β_1-integrins that bound to secreted ECM proteins.

6.3.6 32D-V and 32D-BCR/ABL attachment to FN secreted by BMSC

Subsequently localisations of FN and VCAM were analysed by laser scanning confocal microscopy. A large meshwork of FN fibrils was found on top of the BMSC monolayer. Apparently BMSC secreted FN and assembles it into fibrils (Fig. 68). VCAM-1 expressed by BMSC was homogenously distributed on the cell surface (not shown). After co-culture of BMSC with 32D-V or 32D-BCR/ABL cells, most of attached 32D cells were located at the fibronectin fibrils produced by BMSC. FN was found deposited in the interface between BMSC and 32D cells (Fig. 68, insets). After 3 h of co-culture most 32D-V and 32D-BCR/ABL cells localized with the FN bundles. Thus, confocal microscopy suggested that a major ligand for 32D-V and 32D-BCR/ABL cells was fibronectin expressed by BMSC. However, adhesion of leukemic cells and BMSC certainly involves further molecular interactions than that occurring between $\alpha_5\beta_1$-integrin and fibronectin, for instance integrin-mediated interactions to other ECM proteins secreted by BMSC, such as collagens.

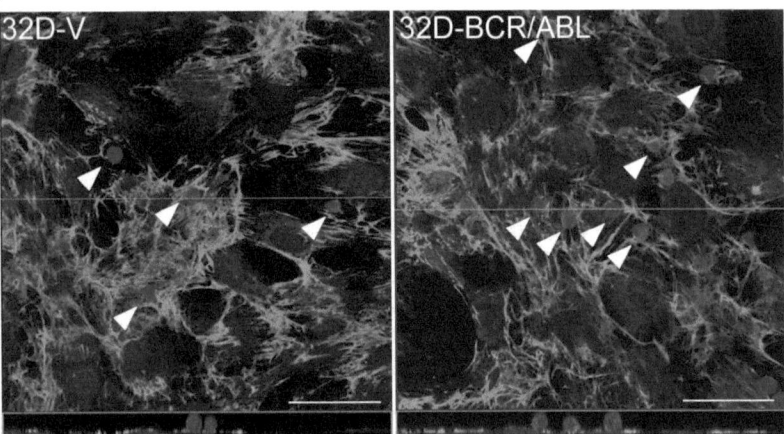

Fig. 68. Confocal imaging of 32D cells on a BMSC monolayer. 32D-V and 32D–BCR/ABL cells were co-cultured for 3h with BMSC. Fixed samples were stained for b1–integrin (red), fibronectin (green) and nuclei (blue). The 3D image was constructed from acquired Z stacks and projected to a X-Y plane. Scale bar corresponds to 50μm. Images were acquired by F. Fierro.

6.4 Conclusions & Outlook

Previous studies have studied the effect of BCR/ABL on cell adhesion. However, in none of these studies cell adhesion was quantitatively measured. Thus, the presented SCFS data provide the first quantitative proof for the enhanced adhesion of BCR/ABL 32D cells to the stromal compartment. Moreover, adhesion to FN and Col was significantly increased. Previous studies reporting enhanced adhesion of BCR/ABL expressing cells contributed the effect of increased adhesion to integrin clustering and cytoskeleton rearrangements without changing β_1-integrin expression[459]. However, above presented flow cytometry data clearly showed enhanced cell surface concentrations of β_1-integrin comprising integrin heterodimers. There might be more than one mechanism by which β_1-integrin mediated adhesion of 32D-BCR/ABL cells is increased. However, since a clear difference of β_1-integrins was observed between 32D-V and –BCR/ABL it might be concluded that different β_1-integrin levels were predominantly responsible for enhanced adhesion of 32D-BCR/ABL. The analysis of single rupture events further suggested that enhanced adhesion was not due to integrin cluster formation. Enhanced integrin clustering in 32D-BCR/ABL cells might have resulted in an increased magnitude of single rupture events compared to 32D-V cells (similar as observed in Chapter 1).

Both enhanced β_1-integrin concentrations and increased adhesion could be reversed by IM in a concentration and time-dependent manner. This confirmed that BCR/ABL tyrosine kinase activity was responsible for the seen effect. Since the inhibition of the BCR/ABL tyrosine kinase by IM occurs within one hour[474] the question arises why such long pre-incubation periods are required to observe an effect of IM on cell adhesion. The prolonged delay until IM incubation effected cell adhesion points to a mechanism that involves secondary signals to reduce cell adhesion. Experiments blocking *de novo* protein synthesis by cycloheximide have shown, that the β_1-integrin concentration on the cell surface required between 4 and 8h to decrease[475]. This might explain why IM treatment of BCR/ABL expressing cells required several hours to affect cell surface β_1-integrin levels.

Taken together, the results show that high BCR/ABL-levels leads to enhanced adhesion of 32D cells to BMSC due to increased β_1-integrin levels. Comparatively high BCR/ABL expression levels have recently been associated with terminal blast crisis of CML and has been detected in the stem cell pool of patients suffering from CML[459, 465]. Since BCR/ABL expression correlates with increased cell adhesion, this increased adhesion may assist leukemic cells to adhere to protected areas of the bone marrow stroma and to become drug resistant. Such an anti-apoptotic effect of the bone marrow microenvironment *in vivo* has been previously demonstrated[453]. In the presented work interactions between progenitor cells and bone marrow cells were dominated by β_1-integrins which is interesting since a crucial role for β_1-integrin in mediating CAM-DR was proposed[452]. Taken together, the presented data suggest that targeting integrin-mediated interaction of CML cells with the bone marrow microenvironment represents a strategy to decrease CAM-DR effects. This might involve antibodies interfering with integrin function or inhibitors acting more downstream. Targeting integrin function and signalling might be interesting especially in cases of imatinib-resistance and might further improve the success of chemotherapeutic drugs.

In the presented work the specific β_1-comprising integrin heterodimers enhancing adhesion of 32D-BCR/ABL to BCSC were not completely characterized. In future work the role of specific integrin heterodimers might be explored by combining RT-PCR, flow cytometry and SCFS experiments. Detailed knowledge about the involved molecules may represent a useful basis to enter the question by which mechanisms BCR/ABL influences β_1-integrin-mediated adhesion.

Final remarks

In the three described projects AFM-based SCFS was applied to investigate integrin-mediated adhesion at the molecular level. In chapter four the kinetics of $\alpha_2\beta_1$-integrin mediated cell adhesion to collagen type I was explored. The high force resolution of the measurements permitted to decipher the contribution of single-integrin interactions in the time-dependent build-up of overall cell adhesion and to elucidate mechanisms of $\alpha_2\beta_1$-integrin regulation. Moreover, dynamic force spectroscopy allowed bond specific parameters to be determined for the $\alpha_2\beta_1$-integrin collagen type I interaction. In chapter five overall cell adhesion of pre-osteoblasts to native and denatured collagen type I was compared. Distinct binding mechanisms and integrins involved in the adhesion process were revealed by using different blocking reagents such as function blocking antibodies and RGD peptide. By combining SCFS with further techniques that permitted analysis of cell spreading, migration, signaling and osteogenic differentiation, the observed differences in cell adhesion could be directly correlated with structural changes of collagen type I that occurred during thermal denaturation at the molecular level. The findings of this project further indicated that the initial phase of cell adhesion, such as assessed by AFM-SCFS, may trigger long-lasting effects on cellular behaviour, such as differentiation. This demonstrates the relevance of studying these initial adhesion events. In chapter six, overall adhesion of BCR/ABL expressing and control cells was compared. By quantifying adhesion in presence of a specific drug, imatinib mesylate, the observed enhanced adhesion of BCR/ABL expressing cells could be correlated with BCR/ABL tyrosine kinase activity. By using a combinatorial approach of SCFS, flow cytometry, RT-PCR and western blots, the molecular mechanisms leading to enhanced adhesion could be investigated. Based on the presented results, it can be concluded that in combination with complementary techniques AFM-SCFS represents a powerful tool to address interesting aspects of cell biological, biophysical and medical relevance.

References

1. Cukierman, E., Pankov, R. & Yamada, K.M. Cell interactions with three-dimensional matrices. *Curr Opin Cell Biol* **14**, 633-639 (2002).
2. Berrier, A.L. & Yamada, K.M. Cell-matrix adhesion. *J Cell Physiol* **213**, 565-573 (2007).
3. Grinnell, F. Fibroblast biology in three-dimensional collagen matrices. *Trends Cell Biol* **13**, 264-269 (2003).
4. Hay, E.D. Biogenesis and organization of extracellular matrix. *Faseb J* **13** Suppl 2, S281-283 (1999).
5. Alberts, B. et al. Molecular Biology of the Cell. (Garland Science, New York; 2002).
6. Pollard, T.D. & Earnshaw, W.C. Cell Biology, Edn. 1st. (Philadelphia; 2004).
7. Wang, X., Bank, R.A., TeKoppele, J.M. & Agrawal, C.M. The role of collagen in determining bone mechanical properties. *J Orthop Res* **19**, 1021-1026 (2001).
8. Katz, J.L. Hard tissue as a composite material. I. Bounds on the elastic behavior. *J Biomech* **4**, 455-473 (1971).
9. Kjaer, M. Role of extracellular matrix in adaptation of tendon and skeletal muscle to mechanical loading. *Physiol Rev* **84**, 649-698 (2004).
10. Holmes, D.F. et al. Corneal collagen fibril structure in three dimensions: Structural insights into fibril assembly, mechanical properties, and tissue organization. *Proc Natl Acad Sci U S A* **98**, 7307-7312 (2001).
11. Vakonakis, I. & Campbell, I.D. Extracellular matrix: from atomic resolution to ultrastructure. *Current Opinion in Cell Biology* **19**, 578-583 (2007).
12. Kadler, K.E., Baldock, C., Bella, J. & Boot-Handford, R.P. Collagens at a glance. *J Cell Sci* **120**, 1955-1958 (2007).
13. van der Rest, M. & Garrone, R. Collagen family of proteins. *Faseb J* **5**, 2814-2823 (1991).
14. Canty, E.G. & Kadler, K.E. Procollagen trafficking, processing and fibrillogenesis. *J Cell Sci* **118**, 1341-1353 (2005).
15. Kadler, K.E., Holmes, D.F., Trotter, J.A. & Chapman, J.A. Collagen fibril formation. *Biochem J* **316** (Pt 1), 1-11 (1996).
16. Orgel, J.P., Irving, T.C., Miller, A. & Wess, T.J. Microfibrillar structure of type I collagen in situ. *Proc Natl Acad Sci U S A* **103**, 9001-9005 (2006).
17. Smith, J.W. Molecular pattern in native collagen. *Nature* **219**, 157-158 (1968).
18. Gross, J. & Schmitt, F.O. The structure of human skin collagen as studied with the electron microscope. *J Exp Med* **88**, 555-568 (1948).
19. Petruska, J.A. & Hodge, A.J. A Subunit Model for the Tropocollagen Macromolecule. *Proc Natl Acad Sci U S A* **51**, 871-876 (1964).
20. Hodge, A.J. & Petruska, J.A. Some recent results on the electron microscopy of tropocollagen structures, Vol. 1. (Academic Press, New York; 1962).
21. Mould, A.P. et al. D-Periodic Assemblies of Type I Procollagen. *J. Mol. Biol.* **211**, 581-594 (1990).
22. Holmes, D.F. & Kadler, K.E. The precision of lateral size control in the assembly of corneal collagen fibrils. *J Mol Biol* **345**, 773-784 (2005).
23. Christiansen, D.L., Huang, E.K. & Silver, F.H. Assembly of type I collagen: fusion of fibril subunits and the influence of fibril diameter on mechanical properties. *Matrix Biology* **19**, 409-420 (2000).
24. Gross, J., Highberger, J.H. & Schmitt, F.O. Some factors involved in the fibrogenesis of collagen in vitro. *Proc Soc Exp Biol Med* **80**, 462-465 (1952).
25. Payne, K.J. & Veis, A. Fourier transform IR spectroscopy of collagen and gelatin solutions: deconvolution of the amide I band for conformational studies. *Biopolymers* **27**, 1749-1760 (1988).
26. Kadler, K.E., Hill, A. & Canty-Laird, E.G. Collagen fibrillogenesis: fibronectin, integrins, and minor collagens as organizers and nucleators. *Curr Opin Cell Biol* **20**, 495-501 (2008).
27. Lee, C.H., Singla, A. & Lee, Y. Biomedical applications of collagen. *Int J Pharm* **221**, 1-22 (2001).
28. Ramshaw, J.A., Peng, Y.Y., Glattauer, V. & Werkmeister, J.A. Collagens as biomaterials. *J Mater Sci Mater Med* (2008).
29. Abraham, L.C., Zuena, E., Perez-Ramirez, B. & Kaplan, D.L. Guide to collagen characterization for biomaterial studies. *Journal of Biomedical Materials Research* Part B: Applied Biomaterials, **264** (2008).
30. Cancedda, R., Dozin, B., Giannoni, P. & Quarto, R. Tissue engineering and cell therapy of cartilage and bone. *Matrix Biology* **22**, 81-91 (2003).
31. Damsky, C., Sutherland, A. & Fisher, S. Extracellular matrix 5: adhesive interactions in early mammalian embryogenesis, implantation, and placentation. *Faseb J* **7**, 1320-1329 (1993).
32. Damsky, C.H. & Ilic, D. Integrin signalling: it's where the action is. *Current Opinion in Cell Biology* **14**, 594-602 (2002).
33. Juliano, R.L. & Haskill, S. Signal transduction from the extracellular matrix. *J Cell Biol* **120**, 577-585 (1993).

34. Bissell, M.J., Hall, H.G. & Parry, G. How does the extracellular matrix direct gene expression? *J Theor Biol* **99**, 31-68 (1982).
35. Adams, J.C. & Watt, F.M. Regulation of development and differentiation by the extracellular matrix. *Development* **117**, 1183-1198 (1993).
36. Meighan, C.M. & Schwarzbauer, J.E. Temporal and spatial regulation of integrins during development. *Curr Opin Cell Biol* **20**, 520-524 (2008).
37. West, C.M. et al. Fibronectin alters the phenotypic properties of cultured chick embryo chondroblasts. *Cell* **17**, 491-501 (1979).
38. Reddi, A.H. & Huggins, C.B. Formation of bone marrow in fibroblast-transformation ossicles. *Proc Natl Acad Sci U S A* **72**, 2212-2216 (1975).
39. Gullberg, D. & Ekblom, P. Extracellular matrix and its receptors during development. *Int J Dev Biol* **39**, 845-854 (1995).
40. Dufour, S., Duband, J.L., Kornblihtt, A.R. & Thiery, J.P. The role of fibronectins in embryonic cell migrations. *Trends Genet* **4**, 198-203 (1988).
41. Darribere, T. et al. Integrins: regulators of embryogenesis. *Biol Cell* **92**, 5-25 (2000).
42. Yost, H.J. Regulation of vertebrate left-right asymmetries by extracellular matrix. *Nature* **357**, 158-161 (1992).
43. Meredith, J.E., Jr., Fazeli, B. & Schwartz, M.A. The extracellular matrix as a cell survival factor. *Mol Biol Cell* **4**, 953-961 (1993).
44. Marastoni, S., Ligresti, G., Lorenzon, E., Colombatti, A. & Mongiat, M. Extracellular matrix: a matter of life and death. *Connect Tissue Res* **49**, 203-206 (2008).
45. Gilmore, A.P. Anoikis. *Cell Death Differ* **12 Suppl 2**, 1473-1477 (2005).
46. Giancotti, F.G. Integrin signaling: specificity and control of cell survival and cell cycle progression. *Curr Opin Cell Biol* **9**, 691-700 (1997).
47. Porter, J.C. & Hogg, N. Integrins take partners: cross-talk between integrins and other membrane receptors. *trends in Cell Biology* **8**, 390-396 (1998).
48. Nathan, C. & Sporn, M. Cytokines in context. *J Cell Biol* **113**, 981-986 (1991).
49. Schwartz, M.A., Schaller, M.D. & Ginsberg, M.H. Integrins: emerging paradigms of signal transduction. *Annu Rev Cell Dev Biol* **11**, 549-599 (1995).
50. Ignotz, R.A. & Massague, J. Cell adhesion protein receptors as targets for transforming growth factor-beta action. *Cell* **51**, 189-197 (1987).
51. Ignotz, R.A., Endo, T. & Massague, J. Regulation of fibronectin and type I collagen mRNA levels by transforming growth factor-beta. *J Biol Chem* **262**, 6443-6446 (1987).
52. Heino, J., Ignotz, R.A., Hemler, M.E., Crouse, C. & Massague, J. Regulation of cell adhesion receptors by transforming growth factor-beta. Concomitant regulation of integrins that share a common beta 1 subunit. *J Biol Chem* **264**, 380-388 (1989).
53. Blatti, S.P., Foster, D.N., Ranganathan, G., Moses, H.L. & Getz, M.J. Induction of fibronectin gene transcription and mRNA is a primary response to growth-factor stimulation of AKR-2B cells. *Proc Natl Acad Sci U S A* **85**, 1119-1123 (1988).
54. Ruoslahti, E. & Pierschbacher, M.D. New perspectives in cell adhesion: RGD and integrins. *Science* **238**, 491-497 (1987).
55. Hynes, R.O. Integrins: a family of cell surface receptors. *Cell* **48**, 549-554 (1987).
56. Bell, G.I. Models for Specific Adhesion of Cells to Cells. *Science* **200**, 618-627 (1978).
57. Greenwalt, D.E. et al. Membrane glycoprotein CD36: a review of its roles in adherence, signal transduction, and transfusion medicine. *Blood* **80**, 1105-1115 (1992).
58. Leitinger, B. & Hohenester, E. Mammalian collagen receptors. *Matrix Biol* **26**, 146-155 (2007).
59. Vuoriluoto, K. et al. Syndecan-1 supports integrin alpha2beta1-mediated adhesion to collagen. *Exp Cell Res* **314**, 3369-3381 (2008).
60. Cohen, M., Joester, D., Geiger, B. & Addadi, L. Spatial and temporal sequence of events in cell adhesion: from molecular recognition to focal adhesion assembly. *Chembiochem* **5**, 1393-1399 (2004).
61. Hadari, Y.R. et al. Galectin-8 binding to integrins inhibits cell adhesion and induces apoptosis. *J Cell Sci* **113 (Pt 13)**, 2385-2397 (2000).
62. Hughes, R.C. Galectins as modulators of cell adhesion. *Biochimie* **83**, 667-676 (2001).
63. Hynes, R.O. Integrins: bidirectional, allosteric signaling machines. *Cell* **110**, 673-687 (2002).
64. Humphries, J.D., Byron, A. & Humphries, M.J. Integrin ligands at a glance. *J Cell Sci* **119**, 3901-3903 (2006).
65. Carrell, N.A., Fitzgerald, L.A., Steiner, B., Erickson, H.P. & Phillips, D.R. Structure of human platelet membrane glycoproteins IIb and IIIa as determined by electron microscopy. *J Biol Chem* **260**, 1743-1749 (1985).
66. Nermut, M.V., Green, N.M., Eason, P., Yamada, S.S. & Yamada, K.M. Electron microscopy and structural model of human fibronectin receptor. *Embo J* **7**, 4093-4099 (1988).
67. Xiong, J.P. et al. Crystal structure of the extracellular segment of integrin alpha Vbeta3. *Science* **294**, 339-345 (2001).
68. Emsley, J., Knight, C.G., Farndale, R.W., Barnes, M.J. & Liddington, R.C. Structural basis of collagen recognition by integrin alpha2beta1. *Cell* **101**, 47-56 (2000).

69. Xiong, J.P. et al. Crystal structure of the extracellular segment of integrin alpha Vbeta3 in complex with an Arg-Gly-Asp ligand. *Science* **296**, 151-155 (2002).
70. Arnaout, M.A., Goodman, S.L. & Xiong, J.P. Structure and mechanics of integrin-based cell adhesion. *Curr Opin Cell Biol* **19**, 495-507 (2007).
71. White, D.J., Puranen, S., Johnson, M.S. & Heino, J. The collagen receptor subfamily of the integrins. *Int J Biochem Cell Biol* **36**, 1405-1410 (2004).
72. Salminen, T.A. et al. Production, crystallization and preliminary X-ray analysis of the human integrin alpha1 I domain. *Acta Crystallogr D Biol Crystallogr* **55**, 1365-1367 (1999).
73. Lee, J.O., Rieu, P., Arnaout, M.A. & Liddington, R. Crystal structure of the A domain from the alpha subunit of integrin CR3 (CD11b/CD18). *Cell* **80**, 631-638 (1995).
74. Qu, A. & Leahy, D.J. Crystal structure of the I-domain from the CD11a/CD18 (LFA-1, alpha L beta 2) integrin. *Proc Natl Acad Sci U S A* **92**, 10277-10281 (1995).
75. Qu, A. & Leahy, D.J. The role of the divalent cation in the structure of the I domain from the CD11a/CD18 integrin. *Structure* **4**, 931-942 (1996).
76. Emsley, J., King, S.L., Bergelson, J.M. & Liddington, R.C. Crystal structure of the I domain from integrin alpha2beta1. *J Biol Chem* **272**, 28512-28517 (1997).
77. Shimaoka, M., Takagi, J. & Springer, T.A. Conformational regulation of integrin structure and function. *Annu Rev Biophys Biomol Struct* **31**, 485-516 (2002).
78. Ginsberg, M.H., Du, X. & Plow, E.F. Inside-out integrin signalling. *Curr Opin Cell Biol* **4**, 766-771 (1992).
79. Coppolino, M.G. & Dedhar, S. Bi-directional signal transduction by integrin receptors. *Int J Biochem Cell Biol* **32**, 171-188 (2000).
80. Pellinen, T. & Ivaska, J. Integrin traffic. *J Cell Sci* **119**, 3723-3731 (2006).
81. Hotchin, N.A., Gandarillas, A. & Watt, F.M. Regulation of cell surface beta 1 integrin levels during keratinocyte terminal differentiation. *J Cell Biol* **128**, 1209-1219 (1995).
82. Zhu, J., Boylan, B., Luo, B.H., Newman, P.J. & Springer, T.A. Tests of the extension and deadbolt models of integrin activation. *J Biol Chem* **282**, 11914-11920 (2007).
83. Calderwood, D.A. Integrin activation. *J Cell Sci* **117**, 657-666 (2004).
84. Hughes, P.E. et al. Breaking the integrin hinge. A defined structural constraint regulates integrin signaling. *J Biol Chem* **271**, 6571-6574 (1996).
85. Ginsberg, M.H. et al. A membrane-distal segment of the integrin alpha IIb cytoplasmic domain regulates integrin activation. *J Biol Chem* **276**, 22514-22521 (2001).
86. Takagi, J., Erickson, H.P. & Springer, T.A. C-terminal opening mimics 'inside-out' activation of integrin alpha5beta1. *Nat Struct Biol* **8**, 412-416 (2001).
87. Tadokoro, S. et al. Talin binding to integrin beta tails: a final common step in integrin activation. *Science* **302**, 103-106 (2003).
88. Calderwood, D.A. Talin controls integrin activation. *Biochem Soc Trans* **32**, 434-437 (2004).
89. Kolanus, W. et al. Alpha L beta 2 integrin/LFA-1 binding to ICAM-1 induced by cytohesin-1, a cytoplasmic regulatory molecule. *Cell* **86**, 233-242 (1996).
90. Kashiwagi, H. et al. Affinity modulation of platelet integrin alphaIIbbeta3 by beta3-endonexin, a selective binding partner of the beta3 integrin cytoplasmic tail. *J Cell Biol* **137**, 1433-1443 (1997).
91. Montanez, E. et al. Kindlin-2 controls bidirectional signaling of integrins. *Genes Dev* **22**, 1325-1330 (2008).
92. O'Toole, T.E. et al. Integrin cytoplasmic domains mediate inside-out signal transduction. *J Cell Biol* **124**, 1047-1059 (1994).
93. Liddington, R.C. & Ginsberg, M.H. Integrin activation takes shape. *J Cell Biol* **158**, 833-839 (2002).
94. Emsley, J., Knight, C.G., Farndale, R.W. & Barnes, M.J. Structure of the integrin alpha2beta1-binding collagen peptide. *J Mol Biol* **335**, 1019-1028 (2004).
95. Luo, B.H. & Springer, T.A. Integrin structures and conformational signaling. *Curr Opin Cell Biol* **18**, 579-586 (2006).
96. Schoenwaelder, S.M. & Burridge, K. Bidirectional signaling between the cytoskeleton and integrins. *Curr Opin Cell Biol* **11**, 274-286 (1999).
97. Miyamoto, S., Akiyama, S.K. & Yamada, K.M. Synergistic roles for receptor occupancy and aggregation in integrin transmembrane function. *Science* **267**, 883-885 (1995).
98. Felsenfeld, D.P., Choquet, D. & Sheetz, M.P. Ligand binding regulates the directed movement of beta1 integrins on fibroblasts. *Nature* **383**, 438-440 (1996).
99. Hato, T., Pampori, N. & Shattil, S.J. Complementary roles for receptor clustering and conformational change in the adhesive and signaling functions of integrin alphaIIb beta3. *J Cell Biol* **141**, 1685-1695 (1998).
100. Connors, W.L. et al. Two synergistic activation mechanisms of alpha2beta1 integrin-mediated collagen binding. *J Biol Chem* **282**, 14675-14683 (2007).
101. Huang, M.M. et al. Adhesive ligand binding to integrin alpha IIb beta 3 stimulates tyrosine phosphorylation of novel protein substrates before phosphorylation of pp125FAK. *J Cell Biol* **122**, 473-483 (1993).
102. Delanoe-Ayari, H., Al Kurdi, R., Vallade, M., Gulino-Debrac, D. & Riveline, D. Membrane and acto-myosin tension promote clustering of adhesion proteins. *Proc Natl Acad Sci U S A* **101**, 2229-2234 (2004).

103. Galbraith, C.G., Yamada, K.M. & Sheetz, M.P. The relationship between force and focal complex development. *J Cell Biol* **159**, 695-705 (2002).
104. Burridge, K., Chrzanowska-Wodnicka, M. & Zhong, C. Focal adhesion assembly. *Trends Cell Biol* **7**, 342-347 (1997).
105. Giancotti, F.G. & Tarone, G. Positional control of cell fate through joint integrin/receptor protein kinase signaling. *Annu Rev Cell Dev Biol* **19**, 173-206 (2003).
106. Moursi, A.M. et al. Fibronectin regulates calvarial osteoblast differentiation. *J Cell Sci* **109** (Pt 6), 1369-1380 (1996).
107. Schwartz, M.A. & Assoian, R.K. Integrins and cell proliferation: regulation of cyclin-dependent kinases via cytoplasmic signaling pathways. *J Cell Sci* **114**, 2553-2560 (2001).
108. Giancotti, F.G. Integrin signaling: specificity and control of cell survival and cell cycle progression. *Curr Opin Cell Biol* **9**, 691-700 (1997).
109. Gilmore, A.P. Anoikis. *Cell Death Differ* **12 Suppl 2**, 1473-1477 (2005).
110. Clark, E.A., King, W.G., Brugge, J.S., Symons, M. & Hynes, R.O. Integrin-mediated signals regulated by members of the rho family of GTPases. *J Cell Biol* **142**, 573-586 (1998).
111. Nobes, C.D. & Hall, A. Rho, rac, and cdc42 GTPases regulate the assembly of multimolecular focal complexes associated with actin stress fibers, lamellipodia, and filopodia. *Cell* **81**, 53-62 (1995).
112. Rottner, K., Hall, A. & Small, J.V. Interplay between Rac and Rho in the control of substrate contact dynamics. *Curr Biol* **9**, 640-648 (1999).
113. Geiger, B. & Bershadsky, A. Assembly and mechanosensory function of focal contacts. *Curr Opin Cell Biol* **13**, 584-592 (2001).
114. Zamir, E. et al. Molecular diversity of cell-matrix adhesions. *J Cell Sci* **112** (Pt 11), 1655-1669 (1999).
115. Zamir, E. & Geiger, B. Components of cell-matrix adhesions. *J Cell Sci* **114**, 3577-3579 (2001).
116. Abercrombie, M. & Dunn, G.A. Adhesions of fibroblasts to substratum during contact inhibition observed by interference reflection microscopy. *Exp Cell Res* **92**, 57-62 (1975).
117. Izzard, C.S. & Lochner, L.R. Cell-to-substrate contacts in living fibroblasts: an interference reflexion study with an evaluation of the technique. *J Cell Sci* **21**, 129-159 (1976).
118. Zamir, E. & Geiger, B. Molecular complexity and dynamics of cell-matrix adhesions. *J. Cell Science* **114**, 3583 (2001).
119. Riveline, D. et al. Focal contacts as mechanosensors: externally applied local mechanical force induces growth of focal contacts by an mDia1-dependent and ROCK-independent mechanism. *J Cell Biol* **153**, 1175-1186 (2001).
120. Woods, A. & Couchman, J.R. Syndecan 4 heparan sulfate proteoglycan is a selectively enriched and widespread focal adhesion component. *Mol. Biol. Cell* **5** (1994).
121. Borowsky, M.L. & Hynes, R.O. Layilin, a novel talin-binding transmembrane protein homologous with C-type lectins, is localized in membrane ruffles. *J Cell Biol* **143**, 429-442 (1998).
122. Tang, H., Kerins, D.M., Hao, Q., Inagami, T. & Vaughan, D.E. The urokinase-type plasminogen activator receptor mediates tyrosine phosphorylation of focal adhesion proteins and activation of mitogen-activated protein kinase in cultured endothelial cells. *J Biol Chem* **273**, 18268-18272 (1998).
123. Wei, Y., Yang, X., Liu, Q., Wilkins, J.A. & Chapman, H.A. A role for caveolin and the urokinase receptor in integrin-mediated adhesion and signaling. *J Cell Biol* **144**, 1285-1294 (1999).
124. Yebra, M., Goretzki, L., Pfeifer, M. & Mueller, B.M. Urokinase-type plasminogen activator binding to its receptor stimulates tumor cell migration by enhancing integrin-mediated signal transduction. *Exp Cell Res* **250**, 231-240 (1999).
125. Gardner, H., Broberg, A., Pozzi, A., Laato, M. & Heino, J. Absence of integrin alpha1beta1 in the mouse causes loss of feedback regulation of collagen synthesis in normal and wounded dermis. *J Cell Sci* **112** (Pt 3), 263-272 (1999).
126. Fassler, R. & Meyer, M. Consequences of lack of beta 1 integrin gene expression in mice. *Genes Dev* **9**, 1896-1908 (1995).
127. Hynes, R.O. Targeted mutations in cell adhesion genes: what have we learned from them? *Dev Biol* **180**, 402-412 (1996).
128. Brakebusch, C., Hirsch, E., Potocnik, A. & Fassler, R. Genetic analysis of beta1 integrin function: confirmed, new and revised roles for a crucial family of cell adhesion molecules. *J Cell Sci* **110** (Pt 23), 2895-2904 (1997).
129. van der Flier, A. & Sonnenberg, A. Function and interactions of integrins. *Cell Tissue Res* **305**, 285-298 (2001).
130. Hogg, N. & Bates, P.A. Genetic analysis of integrin function in man: LAD-1 and other syndromes. *Matrix Biol* **19**, 211-222 (2000).
131. Guo, W. & Giancotti, F.G. Integrin signalling during tumour progression. *Nat Rev Mol Cell Biol* **5**, 816-826 (2004).
132. Tang, C.H. & Wei, Y. The urokinase receptor and integrins in cancer progression. *Cell Mol Life Sci* **65**, 1916-1932 (2008).

133. Ivaska, J. et al. A peptide inhibiting the collagen binding function of integrin alpha2I domain. *J Biol Chem* **274**, 3513-3521 (1999).
134. Kerr, J.R. Cell adhesion molecules in the pathogenesis of and host defence against microbial infection. *Mol Pathol* **52**, 220-230 (1999).
135. Garcia, A.J. & Gallant, N.D. Stick and grip: measurement systems and quantitative analyses of integrin-mediated cell adhesion strength. *Cell Biochem Biophys* **39**, 61-73 (2003).
136. Klebe, R.J., Hall, J.R., Rosenberger, P. & Dickey, W.D. Cell attachment to collagen: the ionic requirements. *Exp Cell Res* **110**, 419-425 (1977).
137. Connors, W.L. & Heino, J. A duplexed microsphere-based cellular adhesion assay. *Anal Biochem* **337**, 246-255 (2005).
138. Kucik, D.F. Measurement of adhesion under flow conditions. *Curr Protoc Cell Biol* **Chapter 9**, Unit 9 6 (2003).
139. Pierres, A. et al. Experimental study of the interaction range and association rate of surface-attached cadherin 11. *Proc Natl Acad Sci U S A* **95**, 9256-9261 (1998).
140. Masson-Gadais, B., Pierres, A., Benoliel, A.M., Bongrand, P. & Lissitzky, J.C. Integrin (alpha) and beta subunit contribution to the kinetic properties of (alpha)2beta1 collagen receptors on human keratinocytes analyzed under hydrodynamic conditions. *J Cell Sci* **112 (Pt 14)**, 2335-2345 (1999).
141. Marshall, B.T. et al. Direct observation of catch bonds involving cell-adhesion molecules. *Nature* **423**, 190-193 (2003).
142. Vitte, J., Benoliel, A.M., Eymeric, P., Bongrand, P. & Pierres, A. Beta-1 integrin-mediated adhesion may be initiated by multiple incomplete bonds, thus accounting for the functional importance of receptor clustering. *Biophys J* **86**, 4059-4074 (2004).
143. Robert, P., Benoliel, A.M., Pierres, A. & Bongrand, P. What is the biological relevance of the specific bond properties revealed by single-molecule studies? *J Mol Recognit* **20**, 432-447 (2007).
144. Lotz, M.M., Burdsal, C.A., Erickson, H.P. & McClay, D.R. Cell adhesion to fibronectin and tenascin: quantitative measurements of initial binding and subsequent strengthening response. *J Cell Biol* **109**, 1795-1805 (1989).
145. McClay, D.R., Wessel, G.M. & Marchase, R.B. Intercellular recognition: quantitation of initial binding events. *Proc Natl Acad Sci U S A* **78**, 4975-4979 (1981).
146. Curtis, A.S. The Mechanism of Adhesion of Cells to Glass. a Study by Interference Reflection Microscopy. *J Cell Biol* **20**, 199-215 (1964).
147. Gupton, S.L. & Waterman-Storer, C.M. Spatiotemporal feedback between actomyosin and focal-adhesion systems optimizes rapid cell migration. *Cell* **125**, 1361-1374 (2006).
148. Owen, G.R., Meredith, D.O., ap Gwynn, I. & Richards, R.G. Focal adhesion quantification - a new assay of material biocompatibility? Review. *Eur Cell Mater* **9**, 85-96; discussion 85-96 (2005).
149. Hunter, A., Archer, C.W., Walker, P.S. & Blunn, G.W. Attachment and proliferation of osteoblasts and fibroblasts on biomaterials for orthopaedic use. *Biomaterials* **16**, 287-295 (1995).
150. Richards, R.G., Owen, G.R., Rahn, B.A. & Gwynn, A.P. A quantitative method of measuring cell-substrate adhesion areas. *Cells and materials* **7**, 15-30 (1997).
151. Charo, I.F., Nannizzi, L., Phillips, D.R., Hsu, M.A. & Scarborough, R.M. Inhibition of fibrinogen binding to GP IIb-IIIa by a GP IIIa peptide. *J Biol Chem* **266**, 1415-1421 (1991).
152. O'Toole, T.E. et al. Integrin cytoplasmic domains mediate inside-out signal transduction. *J Cell Biol* **124**, 1047-1059 (1994).
153. Helenius, J., Heisenberg, C.P., Gaub, H.E. & Muller, D.J. Single-cell force spectroscopy. *J Cell Sci* **121**, 1785-1791 (2008).
154. Evans, E., Ritchie, K. & Merkel, R. Sensitive force technique to probe molecular adhesion and structural linkages at biological interfaces. *Biophys J* **68**, 2580-2587 (1995).
155. Evans, E., Berk, D. & Leung, A. Detachment of agglutinin-bonded red blood cells. *Biophys J* **59**, 838-848 (1991).
156. Zarnitsyna, V.I. et al. Memory in receptor-ligand-mediated cell adhesion. *Proc Natl Acad Sci U S A* **104**, 18037-18042 (2007).
157. Simson, D.A., Ziemann, F., Strigl, M. & Merkel, R. Micropipet-based pico force transducer: in depth analysis and experimental verification. *Biophys J* **74**, 2080-2088 (1998).
158. Leckband, D. & Israelachvili, J. Intermolecular forces in biology. *Q Rev Biophys* **34**, 105-267 (2001).
159. Andersson, M. et al. Using optical tweezers for measuring the interaction forces between human bone cells and implant surfaces: System design and force calibration. *Rev Sci Instrum* **78**, 074302 (2007).
160. Neuman, K.C. & Nagy, A. Single-molecule force spectroscopy: optical tweezers, magnetic tweezers and atomic force microscopy. *Nat methods* **5**, 491-505 (2008).
161. Neuman, K.C., Chadd, E.H., Liou, G.F., Bergman, K. & Block, S.M. Characterization of photodamage to escherichia coli in optical traps. *Biophys J* **77**, 2856-2863 (1999).
162. Liang, H. et al. Wavelength dependence of cell cloning efficiency after optical trapping. *Biophys J* **70**, 1529-1533 (1996).

163. Kollmannsberger, P. & Fabry, B. High-force magnetic tweezers with force feedback for biological applications. *Rev Sci Instrum* **78**, 114301 (2007).
164. Walter, N., Selhuber, C., Kessler, H. & Spatz, J.P. Cellular unbinding forces of initial adhesion processes on nanopatterned surfaces probed with magnetic tweezers. *Nano Lett* **6**, 398-402 (2006).
165. Matthews, B.D. et al. Mechanical properties of individual focal adhesions probed with a magnetic microneedle. *Biochem Biophys Res Commun* **313**, 758-764 (2004).
166. Binnig, G., Quate, C.F. & Gerber, C. Atomic force microscope. *Phys Rev Lett* **56**, 930-933 (1986).
167. Schönenberger, C. & Alvarado, S.F. A differential interferometer for force spectroscopy. *Rev. Sci. Instrum.* **60**, 3131 (1989).
168. Hoogenboom, B.W., Frederix, P.L.T.M., Fotiadis, D., Hug, H.J. & Engel, A. Potential of interferometric cantilever detection and its application for SFM/AFM in liquids. *Nanotechnology* **19**, 6pp (2008).
169. Tortonese, M. Cantilevers and tips for atomic force microscopy. *IEEE Eng Med Biol Mag* **16**, 28-33 (1997).
170. Meyer, G. & Amer, N.M. Novel optical approach to atomic force microscopy. *Appl. Phys. Lett.* **53**, 1045 (1988).
171. Alexander, S. et al. An atomic-resolution atomic-force microscope implemented using an optical lever. *J. appl. phys.* **65**, 164 (1988).
172. Putman, C.A.J., de Grooth, B.G., Van Hulst, N.F. & Greve, J. A detailed analysis of the optical beam deflection technique for use in atomic force microscopy. *J. Appl. Phys.* **72**, 6-12 (1992).
173. Evans, E. Probing the relation between force- lifetime- and chemistry in single molecular bonds. *Annu. Rev. Biophys. Biomol. Struct.* **30**, 105-128 (2001).
174. Florin, E.L., Moy, V.T. & Gaub, H.E. Adhesion forces between individual ligand-receptor pairs. *Science* **264**, 415-417 (1994).
175. Moy, V.T., Florin, E.L. & Gaub, H.E. Intermolecular forces and energies between ligands and receptors. *Science* **266**, 257-259 (1994).
176. Merkel, R., Nassoy, P., Leung, A., Ritchie, K. & Evans, E. Energy landscapes of receptor-ligand bonds explored with dynamic force spectroscopy. *Nature* **397**, 50-53 (1999).
177. Lee, G.U., Chrisey, L.A. & Colton, R.J. Direct measurement of the forces between complementary strands of DNA. *Science* **266**, 771-773 (1994).
178. Dettmann, W. et al. Differences in zero-force and force-driven kinetics of ligand dissociation from beta-galactoside-specific proteins (plant and animal lectins, immunoglobulin G) monitored by plasmon resonance and dynamic single molecule force microscopy. *Arch Biochem Biophys* **383**, 157-170 (2000).
179. Puntheeranurak, T., Wildling, L., Gruber, H.J., Kinne, R.K. & Hinterdorfer, P. Ligands on the string: single-molecule AFM studies on the interaction of antibodies and substrates with the Na+-glucose co-transporter SGLT1 in living cells. *J Cell Sci* **119**, 2960-2967 (2006).
180. Ros, R. et al. Antigen binding forces of individually addressed single-chain Fv antibody molecules. *Proc Natl Acad Sci U S A* **95**, 7402-7405 (1998).
181. Hinterdorfer, P., Baumgartner, W., Gruber, H.J., Schilcher, K. & Schindler, H. Detection and localization of individual antibody-antigen recognition events by atomic force microscopy. *Proc Natl Acad Sci U S A* **93**, 3477-3481 (1996).
182. Tees, D.F., Waugh, R.E. & Hammer, D.A. A microcantilever device to assess the effect of force on the lifetime of selectin-carbohydrate bonds. *Biophys J* **80**, 668-682 (2001).
183. Zhang, X., Bogorin, D.F. & Moy, V.T. Molecular basis of the dynamic strength of the sialyl Lewis X--selectin interaction. *Chemphyschem* **5**, 175-182 (2004).
184. Barsegov, V. & Thirumalai, D. Dynamics of unbinding of cell adhesion molecules: transition from catch to slip bonds. *Proc Natl Acad Sci U S A* **102**, 1835-1839 (2005).
185. Hanley, W. et al. Single molecule characterization of P-selectin/ligand binding. *J Biol Chem* **278**, 10556-10561 (2003).
186. Evans, E., Leung, A., Heinrich, V. & Zhu, C. Mechanical switching and coupling between two dissociation pathways in a P-selectin adhesion bond. *Proc Natl Acad Sci U S A* **101**, 11281-11286 (2004).
187. Marshall, B.T., Sarangapani, K.K., Lou, J., McEver, R.P. & Zhu, C. Force history dependence of receptor-ligand dissociation. *Biophys J* **88**, 1458-1466 (2005).
188. Ratto, T.V., Rudd, R.E., Langry, K.C., Balhorn, R.L. & McElfresh, M.W. Nonlinearly additive forces in multivalent ligand binding to a single protein revealed with force spectroscopy. *Langmuir* **22**, 1749-1757 (2006).
189. Baumgartner, W., Golenhofen, N., Grundhofer, N., Wiegand, J. & Drenckhahn, D. Ca2+ dependency of N-cadherin function probed by laser tweezer and atomic force microscopy. *J Neurosci* **23**, 11008-11014 (2003).
190. Baumgartner, W. et al. Cadherin interaction probed by atomic force microscopy. *Proc Natl Acad Sci U S A* **97**, 4005-4010 (2000).
191. Kokkoli, E., Ochsenhirt, S.E. & Tirrell, M. Collective and single-molecule interactions of alpha5beta1 integrins. *Langmuir* **20**, 2397-2404 (2004).
192. Thie, M. et al. Interactions between trophoblast and uterine epithelium: monitoring of adhesive forces. *Hum Reprod* **13**, 3211-3219 (1998).

193. Li, F., Redick, S.D., Erickson, H.P. & Moy, V.T. Force measurements of the alpha5beta1 integrin-fibronectin interaction. *Biophys J* **84**, 1252-1262 (2003).
194. Puech, P.H. et al. Measuring cell adhesion forces of primary gastrulating cells from zebrafish using atomic force microscopy. *J Cell Sci* **118**, 4199-4206 (2005).
195. Lehenkari, P.P. & Horton, M.A. Single integrin molecule adhesion forces in intact cells measured by atomic force microscopy. *Biochem Biophys Res Commun* **259**, 645-650 (1999).
196. Wojcikiewicz, E.P., Zhang, X., Chen, A. & Moy, V.T. Contributions of molecular binding events and cellular compliance to the modulation of leukocyte adhesion. *J Cell Sci* **116**, 2531-2539 (2003).
197. Benoit, M. Cell adhesion measured by force spectroscopy on living cells. *Methods Cell Biol* **68**, 91-114 (2002).
198. Panorchan, P. et al. Single-molecule analysis of cadherin-mediated cell-cell adhesion. *J Cell Sci* **119**, 66-74 (2006).
199. Evans, E. & Ritchie, K. Dynamic strength of molecular adhesion bonds. *Biophys J* **72**, 1541-1555 (1997).
200. Evans, E.A. & Calderwood, D.A. Forces and bond dynamics in cell adhesion. *Science* **316**, 1148-1153 (2007).
201. Krieg, M., Helenius, J., Heisenberg, C.P. & Muller, D.J. A bond for a lifetime: employing membrane nanotubes from living cells to determine receptor-ligand kinetics. *Angew. Chem. Int. Ed. Engl.* **47**, 9775-9777 (2008).
202. Waugh, R.E. & Hochmuth, R.M. Mechanical equilibrium of thick, hollow, liquid membrane cylinders. *Biophys J* **52**, 391-400 (1987).
203. Derenyi, I., Julicher, F. & Prost, J. Formation and interaction of membrane tubes. *Phys Rev Lett* **88**, 238101 (2002).
204. Hochmuth, R.M. & Marcus, W.D. Membrane tethers formed from blood cells with available area and determination of their adhesion energy. *Biophys J* **82**, 2964-2969 (2002).
205. Sheetz, M.P. Cell control by membrane-cytoskeleton adhesion. *Nat Rev Mol Cell Biol* **2**, 392-396 (2001).
206. Shao, J.Y. & Hochmuth, R.M. Micropipette suction for measuring piconewton forces of adhesion and tether formation from neutrophil membranes. *Biophys J* **71**, 2892-2901 (1996).
207. Davis, D.M. & Sowinski, S. Membrane nanotubes: dynamic long-distance connections between animal cells. *Nat. Rev. Mol. Cell Biol.* **9**, 431-436 (2008).
208. Radmacher, M. Measuring the elastic properties of living cells by the atomic force microscope. *Methods Cell Biol* **68**, 67-90 (2002).
209. Radmacher, M., Fritz, M., Kacher, C.M., Cleveland, J.P. & Hansma, P.K. Measuring the viscoelastic properties of human platelets with the atomic force microscope. *Biophys J* **70**, 556-567 (1996).
210. Franz, C.M., Taubenberger, A., Puech, P.H. & Muller, D.J. Studying integrin-mediated cell adhesion at the single-molecule level using AFM force spectroscopy. *Sci STKE* **2007**, pl5 (2007).
211. Benoit, M., Gabriel, D., Gerisch, G. & Gaub, H.E. Discrete interactions in cell adhesion measured by single-molecule force spectroscopy. *Nat Cell Biol* **2**, 313-317 (2000).
212. Selhuber-Unkel, C., Lopez-Garcia, M., Kessler, H. & Spatz, J.P. Cooperativity in adhesion cluster formation during initial cell adhesion. *Biophys J* **95**, 5424-5431 (2008).
213. Zhang, X., Wojcikiewicz, E.P. & Moy, V.T. Dynamic adhesion of T lymphocytes to endothelial cells revealed by atomic force microscopy. *Exp Biol Med (Maywood)* **231**, 1306-1312 (2006).
214. Zhang, X. et al. Atomic force microscopy measurement of leukocyte-endothelial interaction. *Am J Physiol Heart Circ Physiol* **286**, H359-367 (2004).
215. Zhang, X., Wojcikiewicz, E. & Moy, V.T. Force spectroscopy of the leukocyte function-associated antigen-1/intercellular adhesion molecule-1 interaction. *Biophys J* **83**, 2270-2279 (2002).
216. Zhang, X., Craig, S.E., Kirby, H., Humphries, M.J. & Moy, V.T. Molecular basis for the dynamic strength of the integrin alpha4beta1/VCAM-1 interaction. *Biophys J* **87**, 3470-3478 (2004).
217. Wojcikiewicz, E. et al. LFA-1 binding destabilizes the JAM-A homophilic interaction during leukocyte transmigration. *Biophys J* **96**, 285-293 (2009).
218. Puech, P.H., Poole, K., Knebel, D. & Muller, D.J. A new technical approach to quantify cell adhesion forces by AFM. *Ultramicroscopy* **106**, 637-644 (2006).
219. Taubenberger, A. et al. Revealing early steps of alpha2beta1 integrin-mediated adhesion to collagen type I by using single-cell force spectroscopy. *Mol Biol Cell* **18**, 1634-1644 (2007).
220. Ulrich, F. et al. Wnt11 functions in gastrulation by controlling cell cohesion through Rab5c and E-cadherin. *Dev Cell* **9**, 555-564 (2005).
221. Krieg, M. et al. Tensile forces govern germ-layer organization in zebrafish. *Nat Cell Biol* **10**, 429-436 (2008).
222. Friedrichs, J. et al. Contributions of galectin-3 and -9 to epithelial cell adhesion analyzed by single cell force spectroscopy. *J Biol Chem* **282**, 29375-29383 (2007).
223. Friedrichs, J., Manninen, A., Muller, D.J. & Helenius, J. Galectin-3 regulates integrin alpha2beta1-mediated adhesion to collagen-I and -IV. *J Biol Chem* **283**, 32264-32272 (2008).
224. Tulla, M. et al. TPA primes alpha2beta1 integrins for cell adhesion. *FEBS Lett* **582**, 3520-3524 (2008).
225. Panorchan, P., George, J.P. & Wirtz, D. Probing intercellular interactions between vascular endothelial cadherin pairs at single-molecule resolution and in living cells. *J Mol Biol* **358**, 665-674 (2006).
226. Bajpai, S. et al. {alpha}-Catenin mediates initial E-cadherin-dependent cell-cell recognition and subsequent bond strengthening. *Proc Natl Acad Sci U S A* **105**, 18331-18336 (2008).

227. Hanley, W.D., Wirtz, D. & Konstantopoulos, K. Distinct kinetic and mechanical properties govern selectin-leukocyte interactions. *J Cell Sci* **117**, 2503-2511 (2004).
228. Rosenbluth, M.J., Lam, W.A. & Fletcher, D.A. Force microscopy of nonadherent cells: a comparison of leukemia cell deformability. *Biophys J* **90**, 2994-3003 (2006).
229. Alcaraz, J. et al. Microrheology of human lung epithelial cells measured by atomic force microscopy. *Biophys J* **84**, 2071-2079 (2003).
230. Benoit, M. & Gaub, H.E. Measuring cell adhesion forces with the atomic force microscope at the molecular level. *Cells Tissues Organs* **172**, 174-189 (2002).
231. Lam, W.A., Rosenbluth, M.J. & Fletcher, D.A. Chemotherapy exposure increases leukemia cell stiffness. *Blood* **109**, 3505-3508 (2007).
232. Wu, H.W., Kuhn, T. & Moy, V.T. Mechanical properties of L929 cells measured by atomic force microscopy: effects of anticytoskeletal drugs and membrane crosslinking. *Scanning* **20**, 389-397 (1998).
233. Francius, G., Domenech, O., Mingeot-Leclercq, M.P. & Dufrene, Y.F. Direct observation of Staphylococcus aureus cell wall digestion by lysostaphin. *J Bacteriol* **190**, 7904-7909 (2008).
234. Reich, A., Meurer, M., Eckes, B., Friedrichs, J. & Muller, D.J. Surface morphology and mechanical properties of fibroblasts from scleroderma patients. *J Cell Mol Med* (2008).
235. Cross, S.E., Jin, Y.S., Rao, J. & Gimzewski, J.K. Nanomechanical analysis of cells from cancer patients. *Nat. Nanotechnol.* **2** (2007).
236. Li, Q.S., Lee, G.Y., Ong, C.N. & Lim, C.T. AFM indentation study of breast cancer cells. *Biochem Biophys Res Commun* **374** (2008).
237. Faria, E.C. et al. Measurements of elastic properties of prostate cancer cells using AFM. *Analyst* **133** (2008).
238. Charras, G.T., Lehenkari, P.P. & Horton, M.A. Atomic force microscopy can be used to mechanically stimulate osteoblasts and evaluate cellular strain distribution. *Ultramicroscopy* **86**, 85-95 (2001).
239. Prass, M., Jacobson, K., Mogilner, A. & Radmacher, M. Direct measurement of the lamellipodial protrusive force in a migrating cell. *J Cell Biol* **174**, 767-772 (2006).
240. Brunner, C.A. et al. Cell migration through small gaps. *Eur. Biophys. J.* **35**, 713-719 (2006).
241. Grandbois, M., Dettmann, W., Benoit, M. & Gaub, H.E. Affinity imaging of red blood cells using an atomic force microscope. *J Histochem Cytochem* **48**, 719-724 (2000).
242. Hinterdorfer, P. & Dufrene, Y.F. Detection and localization of single molecular recognition events using atomic force microscopy. *Nat Methods* **3**, 347-355 (2006).
243. Sotres, J. et al. Unbinding molecular recognition force maps of localized single receptor molecules by atomic force microscopy. *Chemphyschem* **9**, 590-599 (2008).
244. Dufrene, Y.F. & Hinterdorfer, P. Recent progress in AFM molecular recognition studies. *Pflugers Arch* **456**, 237-245 (2008).
245. Dufrene, Y.F. Nanoscale exploration of microbial surfaces using the atomic force microscope. *Future Microbiol* **1**, 387-396 (2006).
246. Bell, G.I. Models for Specific Adhesion of Cells to Cells. *Science* **200**, 618-627 (1978).
247. Cohen, M., Joester, D., Geiger, B. & Addadi, L. Spatial and temporal sequence of events in cell adhesion: from molecular recognition to focal adhesion assembly. *Chembiochem* **5**, 1393-1399 (2004).
248. Chen, Y., Lagerholm, B.C., Yang, B. & Jacobson, K. Methods to measure the lateral diffusion of membrane lipids and proteins. *Methods* **39**, 147-153 (2006).
249. Yauch, R.L. et al. Mutational evidence for control of cell adhesion through integrin diffusion/clustering, independent of ligand binding. *J Exp Med* **186**, 1347-1355 (1997).
250. de Brabander, M. et al. Lateral diffusion and retrograde movements of individual cell surface components on single motile cells observed with Nanovid microscopy. *J Cell Biol* **112**, 111-124 (1991).
251. Jacobson, K., Ishihara, A. & Inman, R. Lateral diffusion of proteins in membranes. *Annu Rev Physiol* **49**, 163-175 (1987).
252. Hammer, D.A. & Lauffenburger, D.A. A dynamical model for receptor-mediated cell adhesion to surfaces. *Biophys J* **52**, 475-487 (1987).
253. Evans, E. & Williams, P. Dynamic Force Spectroscopy: I. Single Bonds. (2002).
254. Evans, E. & Ritchie, K. Dynamic strength of molecular adhesion bonds. *Biophys J* **72**, 1541-1555 (1997).
255. Evans, E. Energy landscapes of biomolecular adhesion and receptor anchoring at interfaces explored with dynamic force spectroscopy. *Faraday Discuss.* **111**, 1-16 (1998).
256. Evans, E.A. & Calderwood, D.A. Forces and bond dynamics in cell adhesion. *Science* **316**, 1148-1153 (2007).
257. Marshall, B.T. et al. Direct observation of catch bonds involving cell-adhesion molecules. *Nature* **423**, 190-193 (2003).
258. Zhurkov, S.N. *International Journal of Fracture Mechanics* **1**, 311-323 (1965).
259. Ray, C., Brown, J.R. & Akhremitchev, B.B. Rupture force analysis and the associated systematic errors in force spectroscopy by AFM. *Langmuir* **23**, 6076-6083 (2007).
260. Thomas, W. Catch bonds in adhesion. *Annu Rev Biomed Eng* **10**, 39-57 (2008).
261. Li, F., Redick, S.D., Erickson, H.P. & Moy, V.T. Force measurements of the alpha5beta1 integrin-fibronectin interaction. *Biophys J* **84**, 1252-1262 (2003).

262. Zhang, X., Craig, S.E., Kirby, H., Humphries, M.J. & Moy, V.T. Molecular basis for the dynamic strength of the integrin alpha4beta1/VCAM-1 interaction. *Biophys J* **87**, 3470-3478 (2004).
263. Zhang, X., Wojcikiewicz, E. & Moy, V.T. Force spectroscopy of the leukocyte function-associated antigen-1/intercellular adhesion molecule-1 interaction. *Biophys J* **83**, 2270-2279 (2002).
264. Baumgartner, W. et al. Cadherin interaction probed by atomic force microscopy. *Proc Natl Acad Sci U S A* **97**, 4005-4010 (2000).
265. Baumgartner, W., Golenhofen, N., Grundhofer, N., Wiegand, J. & Drenckhahn, D. Ca2+ dependency of N-cadherin function probed by laser tweezer and atomic force microscopy. *J Neurosci* **23**, 11008-11014 (2003).
266. Marshall, B.T., Sarangapani, K.K., Lou, J., McEver, R.P. & Zhu, C. Force history dependence of receptor-ligand dissociation. *Biophys J* **88**, 1458-1466 (2005).
267. Barsegov, V. & Thirumalai, D. Dynamics of unbinding of cell adhesion molecules: transition from catch to slip bonds. *Proc Natl Acad Sci U S A* **102**, 1835-1839 (2005).
268. Zhang, X., Bogorin, D.F. & Moy, V.T. Molecular basis of the dynamic strength of the sialyl Lewis X--selectin interaction. *Chemphyschem* **5**, 175-182 (2004).
269. Evans, E., Leung, A., Heinrich, V. & Zhu, C. Mechanical switching and coupling between two dissociation pathways in a P-selectin adhesion bond. *Proc Natl Acad Sci U S A* **101**, 11281-11286 (2004).
270. Thomas, W. Catch bonds in adhesion. *Annu. Rev. Biomed. Eng.* **10**, 39-57 (2008).
271. Thomas, W. et al. Catch bond model derived from allostery explains force-activated bacterial adhesion. *Biophys J* **90**, 753-764 (2006).
272. White, D.J., Puranen, S., Johnson, M.S. & Heino, J. The collagen receptor subfamily of the integrins. *Int J Biochem Cell Biol* **36**, 1405-1410 (2004).
273. Heino, J. The collagen family members as cell adhesion proteins. *Bioessays* **29**, 1001-1010 (2007).
274. Wayner, E.A. & Carter, W.G. Identification of multiple cell adhesion receptors for collagen and fibronectin in human fibrosarcoma cells possessing unique alpha and common beta subunits. *J Cell Biol* **105**, 1873-1884 (1987).
275. Gullberg, D. et al. Analysis of alpha 1 beta 1, alpha 2 beta 1 and alpha 3 beta 1 integrins in cell--collagen interactions: identification of conformation dependent alpha 1 beta 1 binding sites in collagen type I. *Embo J* **11**, 3865-3873 (1992).
276. Voigt, S. et al. Distribution and quantification of alpha 1-integrin subunit in rat organs. *Histochem J* **27**, 123-132 (1995).
277. Zutter, M.M. & Santoro, S.A. Widespread histologic distribution of the alpha 2 beta 1 integrin cell-surface collagen receptor. *Am J Pathol* **137**, 113-120 (1990).
278. Santoro, S.A. & Zutter, M.M. The alpha 2 beta 1 integrin: a collagen receptor on platelets and other cells. *Thromb Haemost* **74**, 813-821 (1995).
279. Camper, L., Hellman, U. & Lundgren-Akerlund, E. Isolation, cloning, and sequence analysis of the integrin subunit alpha10, a beta1-associated collagen binding integrin expressed on chondrocytes. *J Biol Chem* **273**, 20383-20389 (1998).
280. Eckes, B. et al. Mechanical tension and integrin alpha 2 beta 1 regulate fibroblast functions. *J Investig Dermatol Symp Proc* **11**, 66-72 (2006).
281. Velling, T., Kusche-Gullberg, M., Sejersen, T. & Gullberg, D. cDNA cloning and chromosomal localization of human alpha(11) integrin. A collagen-binding, I domain-containing, beta(1)-associated integrin alpha-chain present in muscle tissues. *J Biol Chem* **274**, 25735-25742 (1999).
282. Gardner, H., Broberg, A., Pozzi, A., Laato, M. & Heino, J. Absence of integrin alpha1beta1 in the mouse causes loss of feedback regulation of collagen synthesis in normal and wounded dermis. *J Cell Sci* **112** (Pt 3), 263-272 (1999).
283. Pozzi, A., Wary, K.K., Giancotti, F.G. & Gardner, H.A. Integrin alpha1beta1 mediates a unique collagen-dependent proliferation pathway in vivo. *J Cell Biol* **142**, 587-594 (1998).
284. Langholz, O. et al. Collagen and collagenase gene expression in three-dimensional collagen lattices are differentially regulated by alpha 1 beta 1 and alpha 2 beta 1 integrins. *J Cell Biol* **131**, 1903-1915 (1995).
285. Riikonen, T., Vihinen, P., Potila, M., Rettig, W. & Heino, J. Antibody against human alpha 1 beta 1 integrin inhibits HeLa cell adhesion to laminin and to type I, IV, and V collagens. *Biochem Biophys Res Commun* **209**, 205-212 (1995).
286. Tiger, C.F., Fougerousse, F., Grundstrom, G., Velling, T. & Gullberg, D. alpha11beta1 integrin is a receptor for interstitial collagens involved in cell migration and collagen reorganization on mesenchymal nonmuscle cells. *Dev Biol* **237**, 116-129 (2001).
287. Popova, S.N. et al. The mesenchymal alpha11beta1 integrin attenuates PDGF-BB-stimulated chemotaxis of embryonic fibroblasts on collagens. *Dev Biol* **270**, 427-442 (2004).
288. Velling, T., Risteli, J., Wennerberg, K., Mosher, D.F. & Johansson, S. Polymerization of type I and III collagens is dependent on fibronectin and enhanced by integrins alpha 11beta 1 and alpha 2beta 1. *J Biol Chem* **277**, 37377-37381 (2002).
289. Nykvist, P. et al. Distinct recognition of collagen subtypes by alpha(1)beta(1) and alpha(2)beta(1) integrins. Alpha(1)beta(1) mediates cell adhesion to type XIII collagen. *J Biol Chem* **275**, 8255-8261 (2000).

290. Heino, J. The collagen receptor integrins have distinct ligand recognition and signaling functions. *Matrix Biol* **19**, 319-323 (2000).
291. Zhang, W.M. et al. alpha 11beta 1 integrin recognizes the GFOGER sequence in interstitial collagens. *J Biol Chem* **278**, 7270-7277 (2003).
292. Jokinen, J. et al. Integrin-mediated cell adhesion to type I collagen fibrils. *J Biol Chem* **279**, 31956-31963 (2004).
293. Nolte, M. et al. Crystal structure of the alpha1beta1 integrin I-domain: insights into integrin I-domain function. *FEBS Lett* **452**, 379-385 (1999).
294. Kern, A., Eble, J., Golbik, R. & Kuhn, K. Interaction of type IV collagen with the isolated integrins alpha 1 beta 1 and alpha 2 beta 1. *Eur J Biochem* **215**, 151-159 (1993).
295. Kern, A., Briesewitz, R., Bank, I. & Marcantonio, E.E. The role of the I domain in ligand binding of the human integrin alpha 1 beta 1. *J Biol Chem* **269**, 22811-22816 (1994).
296. Tulla, M. et al. Selective binding of collagen subtypes by integrin alpha 1I, alpha 2I, and alpha 10I domains. *J Biol Chem* **276**, 48206-48212 (2001).
297. Knight, C.G. et al. The collagen-binding A-domains of integrins alpha(1)beta(1) and alpha(2)beta(1) recognize the same specific amino acid sequence, GFOGER, in native (triple-helical) collagens. *J Biol Chem* **275**, 35-40 (2000).
298. Emsley, J., Knight, C.G., Farndale, R.W., Barnes, M.J. & Liddington, R.C. Structural basis of collagen recognition by integrin alpha2beta1. *Cell* **101**, 47-56 (2000).
299. Masson-Gadais, B., Pierres, A., Benoliel, A.M., Bongrand, P. & Lissitzky, J.C. Integrin (alpha) and beta subunit contribution to the kinetic properties of (alpha)2beta1 collagen receptors on human keratinocytes analyzed under hydrodynamic conditions. *J Cell Sci* **112 (Pt 14)**, 2335-2345 (1999).
300. Geiger, B., Bershadsky, A., Pankov, R. & Yamada, K.M. Transmembrane crosstalk between the extracellular matrix--cytoskeleton crosstalk. *Nat Rev Mol Cell Biol* **2**, 793-805 (2001).
301. Jiang, F., Horber, H., Howard, J. & Muller, D.J. Assembly of collagen into microribbons: effects of pH and electrolytes. *J Struct Biol* **148**, 268-278 (2004).
302. Cisneros, D.A., Friedrichs, J., Taubenberger, A., Franz, C.M. & Muller, D.J. Creating ultrathin nanoscopic collagen matrices for biological and biotechnological applications. *Small* **3**, 956-963 (2007).
303. Cisneros, D.A., Hung, C., Franz, C.M. & Muller, D.J. Observing growth steps of collagen self-assembly by time-lapse high-resolution atomic force microscopy. *J Struct Biol* **154**, 232-245 (2006).
304. Petruska, J.A. & Hodge, A.J. A Subunit Model for the Tropocollagen Macromolecule. *Proc Natl Acad Sci U S A* **51**, 871-876 (1964).
305. Holmes, D.F., Graham, H.K. & Kadler, K.E. Collagen fibrils forming in developing tendon show an early and abrupt limitation in diameter at the growing tips. *J Mol Biol* **283**, 1049-1058 (1998).
306. Holmes, D.F. et al. Corneal collagen fibril structure in three dimensions: Structural insights into fibril assembly, mechanical properties, and tissue organization. *Proc Natl Acad Sci U S A* **98**, 7307-7312 (2001).
307. Poole, K. et al. Molecular-scale topographic cues induce the orientation and directional movement of fibroblasts on two-dimensional collagen surfaces. *J Mol Biol* **349**, 380-386 (2005).
308. Friedrichs, J., Taubenberger, A., Franz, C.M. & Muller, D.J. Cellular remodelling of individual collagen fibrils visualized by time-lapse AFM. *J Mol Biol* **372**, 594-607 (2007).
309. Tuckwell, D. & Humphries, M.J. Integrin-collagen binding. *Seminars in Cell & Developmental Biology* **7**, 649-657 (1996).
310. Ivaska, J. et al. A peptide inhibiting the collagen binding function of integrin alpha2 I domain. *J Biol Chem* **274**, 3513-3521. (1999).
311. Tulla, M. et al. TPA primes alpha2beta1 integrins for cell adhesion. *FEBS Lett* **582**, 3520-3524 (2008).
312. Vogel, W. et al. Discoidin domain receptor 1 is activated independently of beta(1) integrin. *J Biol Chem* **275**, 5779-5784 (2000).
313. Krensel, K. & Lichtner, R.B. Selective increase of alpha2-integrin sub-unit expression on human carcinoma cells upon EGF-receptor activation. *Int J Cancer* **80**, 546-552 (1999).
314. Evans, E. & Ritchie, K. Dynamic strength of molecular adhesion bonds. *Biophys J* **72**, 1541-1555 (1997).
315. Zhu, J., Boylan, B., Luo, B.H., Newman, P.J. & Springer, T.A. Tests of the extension and deadbolt models of integrin activation. *J Biol Chem* **282**, 11914-11920 (2007).
316. Xu, Y. et al. Multiple binding sites in collagen type I for the integrins alpha1beta1 and alpha2beta1. *J Biol Chem* **275**, 38981-38989 (2000).
317. Li, F., Redick, S.D., Erickson, H.P. & Moy, V.T. Force measurements of the alpha5beta1 integrin-fibronectin interaction. *Biophys J* **84**, 1252-1262 (2003).
318. Zhang, X. et al. Atomic force microscopy measurement of leukocyte-endothelial interaction. *Am J Physiol Heart Circ Physiol* **286**, H359-367 (2004).
319. Wojcikiewicz, E.P., Abdulreda, M.H., Zhang, X. & Moy, V.T. Force spectroscopy of LFA-1 and its ligands, ICAM-1 and ICAM-2. *Biomacromolecules* **7**, 3188-3195 (2006).
320. Zhang, X., Craig, S.E., Kirby, H., Humphries, M.J. & Moy, V.T. Molecular basis for the dynamic strength of the integrin alpha4beta1/VCAM-1 interaction. *Biophys J* **87**, 3470-3478 (2004).

321. Robert, P., Benoliel, A.M., Pierres, A. & Bongrand, P. What is the biological relevance of the specific bond properties revealed by single-molecule studies? *J Mol Recognit* **20**, 432-447 (2007).
322. Bustamante, C., Chemla, Y.R., Forde, N.R. & Izhaky, D. Mechanical processes in biochemistry. *Annu Rev Biochem* **73**, 705-748 (2004).
323. Zhang, X., Wojcikiewicz, E. & Moy, V.T. Force spectroscopy of the leukocyte function-associated antigen-1/intercellular adhesion molecule-1 interaction. *Biophys J* **83**, 2270-2279 (2002).
324. Waugh, R.E. & Hochmuth, R.M. Mechanical equilibrium of thick, hollow, liquid membrane cylinders. *Biophys J* **52**, 391-400 (1987).
325. Bo, L. & Waugh, R.E. Determination of bilayer membrane bending stiffness by tether formation from giant, thin-walled vesicles. *Biophys J* **55**, 509-517 (1989).
326. Derenyi, I., Julicher, F. & Prost, J. Formation and interaction of membrane tubes. *Phys Rev Lett* **88**, 238101 (2002).
327. Hochmuth, R.M. & Marcus, W.D. Membrane tethers formed from blood cells with available area and determination of their adhesion energy. *Biophys J* **82**, 2964-2969 (2002).
328. Sheetz, M.P. Cell control by membrane-cytoskeleton adhesion. *Nat Rev Mol Cell Biol* **2**, 392-396 (2001).
329. Shao, J.Y. & Hochmuth, R.M. Micropipette suction for measuring piconewton forces of adhesion and tether formation from neutrophil membranes. *Biophys J* **71**, 2892-2901 (1996).
330. Krieg, M., Helenius, J., Heisenberg, C.P. & Muller, D.J. A bond for a lifetime: employing membrane nanotubes from living cells to determine receptor-ligand kinetics. *Angew. Chem. Int. Ed. Engl.* **47**, 9775-9777 (2008).
331. Hosu, B.G., Sun, M., Marga, F., Grandbois, M. & Forgacs, G. Eukaryotic membrane tethers revisited using magnetic tweezers. *Phys Biol* **4**, 67-78 (2007).
332. Sun, M. et al. Multiple membrane tethers probed by atomic force microscopy. *Biophys J* **89**, 4320-4329 (2005).
333. Hato, T., Pampori, N. & Shattil, S.J. Complementary roles for receptor clustering and conformational change in the adhesive and signaling functions of integrin alphaIIb beta3. *J Cell Biol* **141**, 1685-1695 (1998).
334. Carman, C.V. & Springer, T.A. Integrin avidity regulation: are changes in affinity and conformation underemphasized? *Curr Opin Cell Biol* **15**, 547-556 (2003).
335. Chen, A. & Moy, V.T. Cross-linking of cell surface receptors enhances cooperativity of molecular adhesion. *Biophys J* **78**, 2814-2820 (2000).
336. Florin, E.L., Moy, V.T. & Gaub, H.E. Adhesion forces between individual ligand-receptor pairs. *Science* **264**, 415-417 (1994).
337. Lehenkari, P.P. & Horton, M.A. Single integrin molecule adhesion forces in intact cells measured by atomic force microscopy. *Biochem Biophys Res Commun* **259**, 645-650 (1999).
338. Ratto, T.V., Rudd, R.E., Langry, K.C., Balhorn, R.L. & McElfresh, M.W. Nonlinearly additive forces in multivalent ligand binding to a single protein revealed with force spectroscopy. *Langmuir* **22**, 1749-1757 (2006).
339. Rottner, K., Hall, A. & Small, J.V. Interplay between Rac and Rho in the control of substrate contact dynamics. *Curr Biol* **9**, 640-648 (1999).
340. Duband, J.L. et al. Fibronectin receptor exhibits high lateral mobility in embryonic locomoting cells but is immobile in focal contacts and fibrillar streaks in stationary cells. *J Cell Biol* **107**, 1385-1396 (1988).
341. Felsenfeld, D.P., Choquet, D. & Sheetz, M.P. Ligand binding regulates the directed movement of beta1 integrins on fibroblasts. *Nature* **383**, 438-440 (1996).
342. Choquet, D., Felsenfeld, D.P. & Sheetz, M.P. Extracellular matrix rigidity causes strengthening of integrin-cytoskeleton linkages. *Cell* **88**, 39-48 (1997).
343. Humphrey, D., Duggan, C., Saha, D., Smith, D. & Kas, J. Active fluidization of polymer networks through molecular motors. *Nature* **416**, 413-416 (2002).
344. Laevsky, G. & Knecht, D.A. Cross-linking of actin filaments by myosin II is a major contributor to cortical integrity and cell motility in restrictive environments. *J Cell Sci* **116**, 3761-3770 (2003).
345. Riento, K. & Ridley, A.J. Rocks: multifunctional kinases in cell behaviour. *Nat Rev Mol Cell Biol* **4**, 446-456 (2003).
346. Chrzanowska-Wodnicka, M. & Burridge, K. Rho-stimulated contractility drives the formation of stress fibers and focal adhesions. *J Cell Biol* **133**, 1403-1415 (1996).
347. Sawada, Y. et al. Force sensing by mechanical extension of the src family kinase substrate p130cas. *Cell* **127**, 1015-1026 (2006).
348. Cohen, M., Joester, D., Geiger, B. & Addadi, L. Spatial and temporal sequence of events in cell adhesion: from molecular recognition to focal adhesion assembly. *Chembiochem* **5**, 1393-1399 (2004).
349. Lotz, M.M., Burdsal, C.A., Erickson, H.P. & McClay, D.R. Cell adhesion to fibronectin and tenascin: quantitative measurements of initial binding and subsequent strengthening response. *J Cell Biol* **109**, 1795-1805 (1989).
350. Gallant, N.D. & Garcia, A.J. Model of integrin-mediated cell adhesion strengthening. *J Biomech* **40**, 1301-1309 (2007).
351. Evans, E.A. & Calderwood, D.A. Forces and bond dynamics in cell adhesion. *Science* **316**, 1148-1153 (2007).

352. Panorchan, P. et al. Single-molecule analysis of cadherin-mediated cell-cell adhesion. *J Cell Sci* **119**, 66-74 (2006).
353. Pierschbacher, M.D. & Ruoslahti, E. Variants of the cell recognition site of fibronectin that retain attachment-promoting activity. *Proc Natl Acad Sci U S A* **81**, 5985-5988 (1984).
354. Pierschbacher, M.D., Hayman, E.G. & Ruoslahti, E. The cell attachment determinant in fibronectin. *J Cell Biochem* **28**, 115-126 (1985).
355. Ruoslahti, E. & Pierschbacher, M.D. New perspectives in cell adhesion: RGD and integrins. *Science* **238**, 491-497 (1987).
356. Bernard, M.P. et al. Structure of a cDNA for the pro alpha 2 chain of human type I procollagen. Comparison with chick cDNA for pro alpha 2(I) identifies structurally conserved features of the protein and the gene. *Biochemistry* **22**, 1139-1145 (1983).
357. Hayman, E.G., Pierschbacher, M.D. & Ruoslahti, E. Detachment of cells from culture substrate by soluble fibronectin peptides. *J Cell Biol* **100**, 1948-1954 (1985).
358. Gullberg, D. et al. Analysis of alpha 1 beta 1, alpha 2 beta 1 and alpha 3 beta 1 integrins in cell--collagen interactions: identification of conformation dependent alpha 1 beta 1 binding sites in collagen type I. *Embo J* **11**, 3865-3873 (1992).
359. White, D.J., Puranen, S., Johnson, M.S. & Heino, J. The collagen receptor subfamily of the integrins. *Int J Biochem Cell Biol* **36**, 1405-1410 (2004).
360. Tulla, M. et al. Selective binding of collagen subtypes by integrin alpha 1I, alpha 2I, and alpha 10I domains. *J Biol Chem* **276**, 48206-48212 (2001).
361. Heino, J. The collagen family members as cell adhesion proteins. *Bioessays* **29**, 1001-1010 (2007).
362. Federman, S., Miller, L.M. & Sagi, I. Following matrix metalloproteinases activity near the cell boundary by infrared micro-spectroscopy. *Matrix Biol* **21**, 567-577 (2002).
363. George, A. & Veis, A. FTIRS in H2O demonstrates that collagen monomers undergo a conformational transition prior to thermal self-assembly in vitro. *Biochemistry* **30**, 2372-2377 (1991).
364. Davis, G.E. Affinity of integrins for damaged extracellular matrix: alpha v beta 3 binds to denatured collagen type I through RGD sites. *Biochem Biophys Res Commun* **182**, 1025-1031 (1992).
365. Yamamoto, M., Yamato, M., Aoyagi, M. & Yamamoto, K. Identification of integrins involved in cell adhesion to native and denatured type I collagens and the phenotypic transition of rabbit arterial smooth muscle cells. *Exp Cell Res* **219**, 249-256 (1995).
366. Davis, G.E., Bayless, K.J., Davis, M.J. & Meininger, G.A. Regulation of tissue injury responses by the exposure of matricryptic sites within extracellular matrix molecules. *Am J Pathol* **156**, 1489-1498 (2000).
367. Humphries, J.D., Byron, A. & Humphries, M.J. Integrin ligands at a glance. *J Cell Sci* **119**, 3901-3903 (2006).
368. Grinnell, F. Fibroblast biology in three-dimensional collagen matrices. *Trends Cell Biol* **13**, 264-269 (2003).
369. Damsky, C., Tremble, P. & Werb, Z. Signal transduction via the fibronectin receptor: do integrins regulate matrix remodeling? *Matrix Suppl* **1**, 184-191 (1992).
370. Damsky, C.H. Extracellular matrix-integrin interactions in osteoblast function and tissue remodeling. *Bone* **25**, 95-96 (1999).
371. Adams, J.C. & Watt, F.M. Regulation of development and differentiation by the extracellular matrix. *Development* **117**, 1183-1198 (1993).
372. Adams, J.C. & Watt, F.M. Changes in keratinocyte adhesion during terminal differentiation: reduction in fibronectin binding precedes alpha 5 beta 1 integrin loss from the cell surface. *Cell* **63**, 425-435 (1990).
373. Abraham, L.C., Dice, J.F., Lee, K. & Kaplan, D.L. Phagocytosis and remodeling of collagen matrices. *Exp Cell Res* **313**, 1045-1055 (2007).
374. Lloyd, A.W. Interfacial bioengineering to enhance surface biocompatibility. *Med Device Technol* **13**, 18-21 (2002).
375. Lee, K.B., Park, S.J., Mirkin, C.A., Smith, J.C. & Mrksich, M. Protein nanoarrays generated by dip-pen nanolithography. *Science* **295**, 1702-1705 (2002).
376. Sato, K. et al. Possible involvement of aminotelopeptide in self-assembly and thermal stability of collagen I as revealed by its removal with proteases. *J Biol Chem* **275**, 25870-25875 (2000).
377. Mauney, J.R. et al. Matrix-mediated retention of in vitro osteogenic differentiation potential and in vivo bone-forming capacity by human adult bone marrow-derived mesenchymal stem cells during ex vivo expansion. *J Biomed Mater Res A* **79**, 464-475 (2006).
378. Mauney, J.R., Kaplan, D.L. & Volloch, V. Matrix-mediated retention of osteogenic differentiation potential by human adult bone marrow stromal cells during ex vivo expansion. *Biomaterials* **25**, 3233-3243 (2004).
379. Mauney, J.R., Volloch, V. & Kaplan, D.L. Matrix-mediated retention of adipogenic differentiation potential by human adult bone marrow-derived mesenchymal stem cells during ex vivo expansion. *Biomaterials* **26**, 6167-6175 (2005).
380. Cisneros, D.A., Friedrichs, J., Taubenberger, A., Franz, C.M. & Muller, D.J. Creating ultrathin nanoscopic collagen matrices for biological and biotechnological applications. *Small* **3**, 956-963 (2007).
381. Holmes, D.F. et al. Corneal collagen fibril structure in three dimensions: Structural insights into fibril assembly, mechanical properties, and tissue organization. *Proc Natl Acad Sci U S A* **98**, 7307-7312 (2001).

382. Werkmeister, J.A., Ramshaw, J.A. & Ellender, G. Characterisation of a monoclonal antibody against native human type I collagen. *Eur J Biochem* **187**, 439-443 (1990).
383. Poole, K. et al. Molecular-scale topographic cues induce the orientation and directional movement of fibroblasts on two-dimensional collagen surfaces. *J Mol Biol* **349**, 380-386 (2005).
384. Zhang, X., Wojcikiewicz, E. & Moy, V.T. Force spectroscopy of the leukocyte function-associated antigen-1/intercellular adhesion molecule-1 interaction. *Biophys J* **83**, 2270-2279 (2002).
385. Helenius, J., Heisenberg, C.P., Gaub, H.E. & Muller, D.J. Single-cell force spectroscopy. *J Cell Sci* **121**, 1785-1791 (2008).
386. Tuckwell, D.S., Ayad, S., Grant, M.E., Takigawa, M. & Humphries, M.J. Conformation dependence of integrin-type II collagen binding. Inability of collagen peptides to support alpha 2 beta 1 binding, and mediation of adhesion to denatured collagen by a novel alpha 5 beta 1-fibronectin bridge. *J Cell Sci* **107 (Pt 4)**, 993-1005 (1994).
387. Pfaff, M. et al. Integrin and Arg-Gly-Asp dependence of cell adhesion to the native and unfolded triple helix of collagen type VI. *Exp Cell Res* **206**, 167-176 (1993).
388. Pacifici, R. et al. Ligand binding to monocyte a5/b1 integrin activates the a2/b1 receptor via the a5 subunit cytoplasmic domain and protein kinase C. *J. Immunol.* **153**, 2222-2233 (1994).
389. Salasznyk, R.M., Williams, W.A., Boskey, A., Batorsky, A. & Plopper, G.E. Adhesion to Vitronectin and Collagen I Promotes Osteogenic Differentiation of Human Mesenchymal Stem Cells. *J Biomed Biotechnol* **2004**, 24-34 (2004).
390. Schwartz, M.A., Schaller, M.D. & Ginsberg, M.H. Integrins: emerging paradigms of signal transduction. *Annu Rev Cell Dev Biol* **11**, 549-599 (1995).
391. Sieg, D.J., Hauck, C.R. & Schlaepfer, D.D. Required role of focal adhesion kinase (FAK) for integrin-stimulated cell migration. *J Cell Sci* **112 (Pt 16)**, 2677-2691 (1999).
392. Salasznyk, R.M., Klees, R.F., Williams, W.A., Boskey, A. & Plopper, G.E. Focal adhesion kinase signaling pathways regulate the osteogenic differentiation of human mesenchymal stem cells. *Exp Cell Res* **313**, 22-37 (2007).
393. Fisher, L.W., Hawkins, G.R., Tuross, N. & Termine, J.D. Purification and partial characterization of small proteoglycans I and II, bone sialoproteins I and II, and osteonectin from the mineral compartment of developing human bone. *J Biol Chem* **262**, 9702-9708 (1987).
394. Hoshi, K., Ejiri, S. & Ozawa, H. Localization alterations of calcium, phosphorus and calcification-related organics such as proteoglycans and alkaline phosphatase during bone calcification. *J Bone Miner Res* **16**, 289-298 (2001).
395. Bellows, C.G., Reimers, S.M. & Heersche, J.N. Expression of mRNAs for type-I collagen, bone sialoprotein, osteocalcin and osteopontin at different stages of osteoblastic differentiation and their regulation by 1,25 dihydroxyvitamin D3. *Cell Tissue Res* **297**, 249-259 (1999).
396. Gregory, C.A., Gunn, W.G., Peister, A. & Prockop, D.J. An Alizarin red-based assay of mineralization by adherent cells in culture: comparison with cetylpyridinium chloride extraction. *Anal Biochem* **329**, 77-84 (2004).
397. Egles, C. et al. Denatured collagen modulates the phenotype of normal and wounded human skin equivalents. *J Invest Dermatol* **128**, 1830-1837 (2008).
398. Moursi, A.M. et al. Fibronectin regulates calvarial osteoblast differentiation. *J Cell Sci* **109 (Pt 6)**, 1369-1380 (1996).
399. Stephansson, S.N., Byers, B.A. & Garcia, A.J. Enhanced expression of the osteoblastic phenotype on substrates that modulate fibronectin conformation and integrin receptor binding. *Biomaterials* **23**, 2527-2534 (2002).
400. Moursi, A.M., Globus, R.K. & Damsky, C.H. Interactions between integrin receptors and fibronectin are required for calvarial osteoblast differentiation in vitro. *J Cell Sci* **110 (Pt 18)**, 2187-2196 (1997).
401. Garcia, A.J., Vega, M.D. & Boettiger, D. Modulation of cell proliferation and differentiation through substrate-dependent changes in fibronectin conformation. *Mol Biol Cell* **10**, 785-798 (1999).
402. Keselowsky, B.G., Collard, D.M. & Garcia, A.J. Integrin binding specificity regulates biomaterial surface chemistry effects on cell differentiation. *Proc Natl Acad Sci U S A* **102**, 5953-5957 (2005).
403. Bissell, M.J., Hall, H.G. & Parry, G. How does the extracellular matrix direct gene expression? *J Theor Biol* **99**, 31-68 (1982).
404. Juliano, R.L. & Haskill, S. Signal transduction from the extracellular matrix. *J Cell Biol* **120**, 577-585 (1993).
405. Vogel, V. & Baneyx, G. The tissue engineering puzzle: a molecular perspective. *Annu Rev Biomed Eng* **5**, 441-463 (2003).
406. Schwartz, M.A. & DeSimone, D.W. Cell adhesion receptors in mechanotransduction. *Curr Opin Cell Biol* **20**, 551-556 (2008).
407. Engler, A.J., Sweeney, H.L., Discher, D.E. & Schwarzbauer, J.E. Extracellular matrix elasticity directs stem cell differentiation. *J Musculoskelet Neuronal Interact* **7**, 335 (2007).
408. Engler, A.J., Sen, S., Sweeney, H.L. & Discher, D.E. Matrix elasticity directs stem cell lineage specification. *Cell* **126**, 677-689 (2006).

409. Jokinen, J. et al. Integrin-mediated cell adhesion to type I collagen fibrils. *J Biol Chem* **279**, 31956-31963 (2004).
410. Comisar, W.A., Kazmers, N.H., Mooney, D.J. & Linderman, J.J. Engineering RGD nanopatterned hydrogels to control preosteoblast behaviour: A combined computational and experimental approach. *Biomaterials* **28**, 4409-4417 (2007).
411. Arnold, M., Calvalcanti-Adam, E.A. & Spatz, J.P. xxx. *ChemPhysChem* **5**, 383-388 (200).
412. Virchow, R. Weisses Blut. *Frorieps Notitzen* **36**, 151-156 (1845).
413. Wertheim, J.A., Miller, J.P., Xu, L., He, Y. & Pear, W.S. The biology of chronic myelogenous leukemia:mouse models and cell adhesion. *Oncogene* **21**, 8612-8628 (2002).
414. Pasternak, G., Hochhaus, A., Schultheis, B. & Hehlmann, R. Chronic myelogenous leukemia: molecular and cellular aspects. *J Cancer Res Clin Oncol* **124**, 643-660 (1998).
415. Nowell, P.C. & Hungerford, D.A. A minute chromosome in human granulocytic leukemia. *Science* **132**, 1407 (1960).
416. Kurzrock, R. et al. BCR rearrangement-negative chronic myelogenous leukemia revisited. *J Clin Oncol* **19**, 2915-2926 (2001).
417. Ren, R. Mechanisms of BCR-ABL in the pathogenesis of hcronic myelogenous leukemia. *Nat Rev Cancer* **5**, 172-183 (2005).
418. Rowley, J.D. Letter: A new consistent chromosomal abnormality in chronic myelogenous leukaemia identified by quinacrine fluorescence and Giemsa staining. *Nature* **243**, 290-293 (1973).
419. Heisterkamp, N. et al. Localization of the c-abl oncogene adjacent to a translocation break point in chronic myelocytic leukaemia. *Nature* **306**, 239-242 (1983).
420. de Klein, A., Van Kessel, A.G. & Grosveld, G. A cellular oncogene is translocated to the Philadelphia chromosome in chronic myelocytic leukemia. *Nature* **300**, 765-767 (1982).
421. Shtivelman, E., Lifshitz, B., Gale, R.P. & Canaani, E. Fused transcript of abl and bcr genes in chronic myelogenous leukaemia. *Nature* **315**, 550-554 (1985).
422. Ben-Neriah, Y., Daley, G.Q., Mes-Masson, A.-M., Witte, O.N. & Baltimore, D. The chronic myelogenous leukemia-specifc P210 protein is the product of the bcr/abl hybrid gene. *Science* **233**, 212-214 (1986).
423. Daley, G.Q. & Baltimore, D. *PNAS* **85**, 9312-9316 (1988).
424. McLaughlin, J., Chianese, E. & Witte, O.N. *PNAS* **84**, 6558-6562. (1987).
425. Hariharan, I.K. et al. *Mol. Cell Biol.* **9**, 2798-2805 (1989).
426. Pear, W.S. & al., e. Efficient and rapid induction of a chronic myelogenous leukemia-like myeloproliferative disease in mice receiving P210 bcr/abl-transduced bone marrow. *Blood* **92**, 3780-3792 (1998).
427. Daley, G.Q., Van Etten, R.A. & Baltimore, D. Induction of chronic myelogenous leukemia in mice by the P210bcr/abl gene of the Philadelphia chromosome. *Science* **247**, 824-830 (1990).
428. Lin, J. & Arlinghaus, R. Activated c-Abl tyrosine kinase in malignant solid tumors. *Oncogene* **27**, 4385-4391 (2008).
429. Woodring, P.J., Hunter, T. & Wang, J.Y. Regulation of F-actin-dependent processes by the Abl family of tyrosine kinases. *J Cell Sci* **116**, 2613-2626 (2003).
430. Plattner, R. & Pendergast, A.M. Activation and signaling of the Abl tyrosine kinase: bidirectional link with phosphoinositide signaling. *Cell Cycle* **2**, 273-274 (2003).
431. Plattner, R. et al. A new link between the c-Abl tyrosine kinase and phosphoinositide signalling through PLC-gamma1. *Nat Cell Biol* **5**, 309-319 (2003).
432. Chu, S., Li, L., Singh, H. & Bhatia, R. BCR-tyrosine 177 plays an essential role in Ras and Akt activation and in human hematopoietic progenitor transformation in chronic myelogenous leukemia. *Cancer Res* **67**, 7045-7053 (2007).
433. Kharas, M.G. & Fruman, D.A. ABL oncogenes and phosphoinositide 3-kinase: mechanism of activation and downstream effectors. *Cancer Res* **65**, 2047-2053 (2005).
434. Chai, S.K., Nichols, G.L. & Rothman, P. Constitutive activation of JAKs and STATs in BCR-Abl-expressing cell lines and peripheral blood cells derived from leukemic patients. *J Immunol* **159**, 4720-4728 (1997).
435. Steelman, L.S. et al. JAK/STAT, Raf/MEK/ERK, PI3K/Akt and BCR-ABL in cell cycle progression and leukemogenesis. *Leukemia* **18**, 189-218 (2004).
436. Jin, A. et al. BCR/ABL and IL-3 activate Rap1 to stimulate the B-Raf/MEK/Erk and Akt signaling pathways and to regulate proliferation, apoptosis, and adhesion. *Oncogene* **25**, 4332-4340 (2006).
437. Chu, S., Holtz, M., Gupta, M. & Bhatia, R. BCR/ABL kinase inhibition by imatinib mesylate enhances MAP kinase activity in chronic myelogenous leukemia CD34+ cells. *Blood* **103**, 3167-3174 (2004).
438. Kim, J.H. et al. Activation of the PI3K/mTOR pathway by BCR-ABL contributes to increased production of reactive oxygen species. *Blood* **105**, 1717-1723 (2005).
439. Hall, A.G. & Irving, J. New drugs, new drug resistance mechanisms. *Expert Rev Anticancer Ther* **2**, 239-240 (2002).
440. Srinivasan, D. & Plattner, R. Activation of Abl tyrosine kinases promotes invasion of aggressive breast cancer cells. *Cancer Res* **66**, 5648-5655 (2006).

441. Deininger, M.W., Goldman, J.M. & Melo, J.V. The molecular biology of chronic myeloid leukemia. *Blood* **96**, 3343-3356 (2000).
442. Hehlmann, R., Hochhaus, A. & Baccarani, M. Chronic myeloid leukaemia. *Lancet* **370**, 342-350 (2007).
443. Druker, B.J. & Lydon, N.B. Lessons learned from the development of an abl tyrosine kinase inhibitor for chronic myelogenous leukemia. *J Clin Invest* **105**, 3-7 (2000).
444. Druker, B.J. et al. Efficacy and safety of a specific inhibitor of the BCR-ABL tyrosine kinase in chronic myeloid leukemia. *N Engl J Med* **344**, 1031-1037 (2001).
445. Druker, B.J. Translation of the Philadelphia chromosome into therapy for CML. *Blood* **112**, 4808-4817 (2008).
446. Kantarjian, H., Giles, F., Quintás-Cardama, A. & Cortes, J. Important therapeutic targets in chronic myelogenous leukemia. *Clin Cancer Res* **13**, 1089-1097 (2007).
447. Heaney, N.B. & Holyoake, T.L. Therapeutic targets in chronic myeloid leukaemia. *Hematol Oncol* **25**, 66-75 (2007).
448. McCubrey, J.A. et al. Targeting the Raf/MEK/ERK pathway with small-molecule inhibitors. *Curr Opin Investig Drugs* **9**, 614-630 (2008).
449. McCubrey, J.A. et al. Targeting survival cascades induced by activation of Ras/Raf/MEK/ERK, PI3K/PTEN/Akt/mTOR and Jak/STAT pathways for effective leukemia therapy. *Leukemia* **22**, 708-722 (2008).
450. Li, Z. & Li, L. Understanding hematopoietic stem-cell microenvironment. *Trends Biochem. Sci.* **31**, 589-595 (2006).
451. Rizo, A., Vallenga, E., de Haan, G. & Schuringa, J.J. *Hum. Mol. Genet.* **15**, R210-R219. (2006).
452. Hazlehurst, L.A., Damiano, J.S., Buyuksal, I., Pledger, W.J. & Dalton, W.S. Adhesion to fibronectin via beta1 integrins regulates p27kip1 levels and contributes to cell adhesion mediated drug resistance (CAM-DR). *Oncogene* **19**, 4319-4327 (2000).
453. Matsunaga, T. et al. Interaction between leukemic-cell VLA-4 and stromal fibronectin is a decisive factor for minimal residual disease of acute myelogenous leukemia. *Nat Med* **9**, 1158-1165 (2003).
454. Eaves, A.C., Cashman, J.D., Gaboury, L.A., Kalousek, D.K. & Eaves, C.J. Unregulated proliferation of primitive chronic myeloid leukemia progenitors in the presence of normal marrow adherent cells. *Proc Natl Acad Sci U S A* **83**, 5306-5310 (1986).
455. Gordon, M.Y., Dowding, C.R., Riley, G.P., Goldman, J.M. & Greaves, M.F. Altered adhesive interactions with marrow stroma of haematopoietic progenitor cells in chronic myeloid leukaemia. *Nature* **328**, 342-344 (1987).
456. Kramer, A. et al. Adhesion to fibronectin stimulates proliferation of wild-type and bcr/abl-transfected murine hematopoietic cells. *Proc Natl Acad Sci U S A* **96**, 2087-2092 (1999).
457. Wertheim, J.A. et al. Localization of BCR-ABL to F-actin regulates cell adhesion but does not attenuate CML development. *Blood* **102**, 2220-2228 (2003).
458. Bhatia, R., Munthe, H.A. & Forman, S.J. Abnormal growth factor modulation of beta1-integrin-mediated adhesion in chronic myelogenous leukaemia haematopoietic progenitors. *Br J Haematol* **115**, 845-853 (2001).
459. Bazzoni, G., Carlesso, N., Griffin, J.D. & Hemler, M.E. Bcr/Abl expression stimulates integrin function in hematopoietic cell lines. *J Clin Invest* **98**, 521-528 (1996).
460. Ramaraj, P. et al. Effect of mutational inactivation of tyrosine kinase activity on BCR/ABL-induced abnormalities in cell growth and adhesion in human hematopoietic progenitors. *Cancer Res* **64**, 5322-5331 (2004).
461. Zhao, R.C., Jiang, Y. & Verfaillie, C.M. A model of human p210(bcr/ABL)-mediated chronic myelogenous leukemia by transduction of primary normal human CD34(+) cells with a BCR/ABL-containing retroviral vector. *Blood* **97**, 2406-2412 (2001).
462. Bhatia, R. & Verfaillie, C.M. Inhibition of BCR-ABL expression with antisense oligodeoxynucleotides restores beta1 integrin-mediated adhesion and proliferation inhibition in chronic myelogenous leukemia hematopoietic progenitors. *Blood* **91**, 3414-3422 (1998).
463. Salesse, S. & Verfaillie, C.M. Mechanisms underlying abnormal trafficking and expansion of malignant progenitors in CML: BCR/ABL-induced defects in integrin function in CML. *Oncogene* **21**, 8605-8611 (2002).
464. Barnes, D.J., Schultheis, B., Adedeji, S. & Melo, J.V. Dose-dependent effects of Bcr-Abl in cell line models of different stages of chronic myeloid leukaemia. *Oncogene* **24**, 6432-6440 (2005).
465. Jamieson, C.H. et al. Granulocyte-macrophage progenitors as candidate leukemic stem cells in blast-crisis CML. *N Engl J Med* **351**, 657-667 (2004).
466. Wertheim, J.A. et al. BCR-ABL-induced adhesion defects are tyrosine kinase-independent. *Blood* **99**, 4122-4130 (2002).
467. Peng, B. et al. Pharmacokinetics and pharmacodynamics of imatinib in a phase I trial with chronic myeloid leukemia patients. *J Clin Oncol* **22**, 935-942 (2004).
468. Schleyer, E. et al. Liquid chromatographic method for detection and quantitation of STI-571 and its main metabolite N-desmethyl-STI in plasma, urine, cerebrospinal fluid, culture medium and cell preparations. *J Chromatogr B Analyt Technol Biomed Life Sci* **799**, 23-36 (2004).
469. Springer, T.A. Adhesion receptors of the immune system. *Nature* **346**, 425-434 (1990).

470. Mould, A.P., Akiyama, S.K. & Humphries, M.J. Regulation of integrin alpha 5 beta 1-fibronectin interactions by divalent cations. Evidence for distinct classes of binding sites for Mn2+, Mg2+, and Ca2+. *J Biol Chem* **270**, 26270-26277 (1995).
471. Kirchhofer, D., Grzesiak, J. & Pierschbacher, M.D. Calcium as a potential physiological regulator of integrin-mediated cell adhesion. *J Biol Chem* **266**, 4471-4477 (1991).
472. Heino, J., Ignotz, R.A., Hemler, M.E., Crouse, C. & Massague, J. Regulation of cell adhesion receptors by transforming growth factor-beta. *Journal of Biological Chemistry* **264**, 380-388 (1998).
473. Humphries, J.D., Byron, A. & Humphries, M.J. Integrin ligands at a glance. *J Cell Sci* **111**, 3901-3903 (2006).
474. Gambacorti-Passerini, C. et al. Inhibition of the ABL kinase activity blocks the proliferation of BCR/ABL+ leukemic cells and induces apoptosis. *Blood Cells Mol Dis* **23**, 380-394 (1997).
475. Dalton, S.L., Scharf, E., Briesewitz, R., Marcantonio, E.E. & Assoian, R.K. Cell adhesion to extracellular matrix regulates the life cycle of integrins. *Mol Biol Cell* **6**, 1781-1791 (1995).

Index of figures and tables

Index of figures

		Page
Fig. 1.	Chicken fibro-blast in the connective tissue of the skin	9
Fig. 2.	Axial structure of D-periodic collagen type I fibrils	12
Fig. 3.	Overview of integrin heterodimers	15
Fig. 4.	Combinations of integrin-ECM interactions	15
Fig. 5.	Molecular model of integrin $\alpha 2\beta 1$ ectodomain in its extended Conformation	16
Fig. 6.	Structure of the $\alpha 2$I-domain in complex with a collagen like peptide	17
Fig. 7.	Different mechanisms to regulate integrin mediated adhesion	18
Fig. 8.	Integrin structural changes during activation	19
Fig. 9.	Integrin-cytoskeleton interactions	21
Fig. 10:	Integrin inside-out and outside-in signaling	22
Fig. 11.	Focal adhesions visualized by fluorescence microscopy	23
Fig. 12.	AFM cantilever and principle of force detection	33
Fig. 13.	AFM force spectroscopy	34
Fig. 14.	Possible SCFS experimental setups to measure cell interactions with adhesive substrates.	35
Fig. 15.	Attaching a living cell onto the AFM cantilever	36
Fig. 16.	Monitoring the force signal during a F-D cycle	37
Fig. 17.	F-D curve and extracted information	37
Fig. 18.	Schematic representation of the cell detachment process	38
Fig. 19.	Sketch illustrating the different events causing j and t-like events	40
Fig. 20	Image of a membrane tether pulled from a embryonic zebrafish cell	41
Fig. 21.	F-D curves in presence of Cytochalasin D	42
Fig. 22.	Alternative SCFS setups	44
Fig. 23.	Sketch illustrating the contact zone established between a cell and a ligand-coated surfaces during AFM SCFS	48
Fig. 24.	Schematic representation of receptor-ligand bond dissociation	50
Fig. 25	Influence of an external force on the energy barrier separating the bond and the unbound state	51
Fig. 26.	Representative F-D curve recorded in AFM-SCFS experiments	52
Fig. 27.	Schematic representation showing the effects of force on average bond Lifetime and binding strength	54
Fig. 28 .	AFM topographs of two-dimensional collagen I matrices	61
Fig. 29.	Integrin $\alpha 2$ expression in Saos-WT/-A2 and CHO-WT/-A2 cells	62
Fig. 30.	Effect of trypsin treatment on $\alpha 2$- and $\beta 1$-integrins	63
Fig. 31.	Spreading of Saos-WT and –A2 cells on Col matrices	64
Fig. 32.	Spreading of CHO-WT and –A2 cells on Col matrices	64
Fig. 33:	Quantifying adhesion of Saos-A2 and -WT cells to Col	65
Fig. 34.	Quantifying cell adhesion of CHO-WT and -A2 cells to Col	66
Fig. 35.	Analysis of CHO cell diameter and $\alpha 2$-integrin expression	68
Fig. 36.	Detecting single $\alpha 2\beta 1$- integrin Col interactions	69
Fig. 37.	Analyzing $\alpha 2\beta 1$-integrin-collagen interactions by dynamic force spectroscopy	70

		Page
Fig. 38.	j- and t- events at different pulling velocities	73
Fig. 39.	Monitoring cell shape during contact	74
Fig. 40.	Detachment forces of CHO-A2 and -WT cells for 0 - 30 sec contact	75
Fig. 41.	Dependence of detachment forces on contact time	75
Fig. 42.	Example F-D curves displaying increasing detachment forces for prolonged contact	76
Fig. 43.	Separating high- and low- adhesion cells	77
Fig. 44.	Single rupture events for high- and low-adhesion	79
Fig. 45.	Influence of inhibitors of actomyosin contractility on cell adhesion	81
Fig. 46.	Confocal microscopy images of the contact zone of a YFP-paxillin expressing CHO-A2 cells and Col during SCFS	83
Fig. 47.	Overview about the sequential build-up of $\alpha 2\beta 1$-integrin-collagen type I bonds	85
Fig. 48.	Sketch illustrating the effect of thermal denaturation on collagen type I structure	88
Fig. 49.	Characterization of two-dimensional collagen type I matrices	91
Fig. 50.	Characterisation of Col and pdCol by immunofluorescence microscopy	92
Fig. 51.	Characterizing mechanical properties of Col and pdCol	93
Fig. 52.	Cellular behaviour on Col/pdCol matrices	94
Fig. 53.	Quantification of cell adhesion to Col/pdCol matrices by SCFS	95
Fig. 54.	Flow cytometry analysis of cell surface associated integrins	96
Fig. 55.	Quantifying the effect of integrin blocking on cell adhesion to Col and pdCol	97
Fig. 56.	FAK tyr397 phosphorylation on Col/pdCol	99
Fig. 57.	Cell proliferation and matrix mineralization	100
Fig. 58.	Reciprocal interactions between cells and ECM	103
Fig. 59.	Signalling pathways affected by BCR/ABL	108
Fig. 60.	Mechanism of IM	109
Fig. 61.	Quantification of cell-cell adhesion using washing assays	111
Fig. 62.	Experimental setup to quantify cell-cell adhesion by SCFS	112
Fig. 63.	Quantification of cell-cell adhesion by SCFS	112
Fig. 64.	Quantifying adhesion of 32D cells to ECM proteins	115
Fig. 65.	Quantifying adhesion of 32D cells to BMSC blocking $\beta 1$-integrins	116
Fig. 66.	$\beta 1$-integrin and gene expression in 32D cells	118
Fig. 67.	Effect of BCR/ABL on 32D cell surface concentrations of integrins	119
Fig. 68.	Confocal imaging of 32D cells on a BMSC monolayer	120

Index of tables

Table 1.	Overview about ECM macromolecules	10
Table 2.	Overview about cell adhesion assays	32
Table 3.	Percentage of single rupture events ("j") <73 pN detected for different experimental conditions (inhibitors, contact times)	82
Table 4.	Analysis of single force jumps for cell-cell and cell-ECM measurements	114

Appendix

Contents of Appendix

			Page
A	**Supplementary information**		**147**
	A1	Supplementary information for chapter 2	147
	A2	Supplementary information for chapter 4	147
		A2.1 Testing the effect of repeated F-D cycles	*147*
		A2.2 Testing the mechanical stability of Col	*148*
		A2.3 Statistics of SCFS data	*149*
	A3	Supplementary information for chapter 5	150
B	**Experimental procedures**		**152**
	B1	AFM-SCFS	152
		B1.1 Experiments	*152*
		B1.2 Data analysis- Overall cell adhesion	*157*
		B1.3 Modified setup for single-molecule measurements	159
	B2	Preparation of adhesive ECM substrates	163
	B3	AFM imaging	164
	B4	Analyzing mechanical properties of collagen type I matrices	165
	B5	Immunostaining of Col/pdCol matrices (mica)	165
	B6	Immunostaining of cells	166
	B7	Analysis of cell migration and spreading	166
	B8	FAK phosphorylation at tyr 397	167
	B9	Washing assays	167
	B10	Cell proliferation assay	167
	B11	Matrix mineralization- Alizarin red stain	168
	B12	Flow cytometry	168
	B13	SDS-PAGE and western blots	169
	B14	Reverse transcription and quantitative real-time PCR	169
	B15	Statistical analysis	169
C	**Cell culture**		**170**
D	**Materials**		**173**
	D1	Chemicals and media	173
	D2	Plastic equipment	174
	D3	AFM cantilevers	174
	D4	ECM proteins	174
	D5	Proteins for cantilever functionalization	174
	D6	Antibodies and blocking peptides	174

A Supplementary information

A1 Supplementary information for chapter 2

Covariance between parameters extracted from F-D curves

Covariances for detachment force (F), work (W), separation distance d and cell elasticity were determined using *Matlab* (*MathWorks*) for a dataset recorded for 32D-V cells. E was determined using the area under the approach curve in the contact regime according to[*]. F and W are clearly related (cov=0.67). Furthermore, there is a higher covariance for W-E (cov=0.33) compared to F-E (cov=0.08). This indicates that W is more influenced by cell elasticity than F. Similarly, separation distance d has more impact on W than on F. Membrane nanotubes that increase the separation distance, contribute substantially to increase W.

	F	W	d	E
F	1	0.67	0.15	0.08
W	0.67	1	0.58	0.33
d	0.15	0.58	1	0.26
E	0.08	0.33	0.26	1

Table S1. Covariance between parameters extracted form F-D curves. For a dataset recorded with 32D cells (5sec contact, 75F-D cycles) the covariance between detachment work W, force F, elasticity E and separation distance d was calculated.

A2 Supplementary information for chapter 4

A2.1 Testing the effect of repeated F-D cycles.

To test if repeating F-D cycles had an effect on subsequent F-D cycles, detachment forces of five repeatedly recorded F-D cycles were compared (Fig. S1). Since detachment forces did not significantly change for the tested contact times, it was concluded that the contact history did not bias the measurements.

---

[*] Rosenbluth, M.J., Lam, W.A. & Fletcher, D.A. Force microscopy of nonadherent cells: a comparison of leukemia cell deformability. *Biophys J* **90**, 2994-3003 (2006).

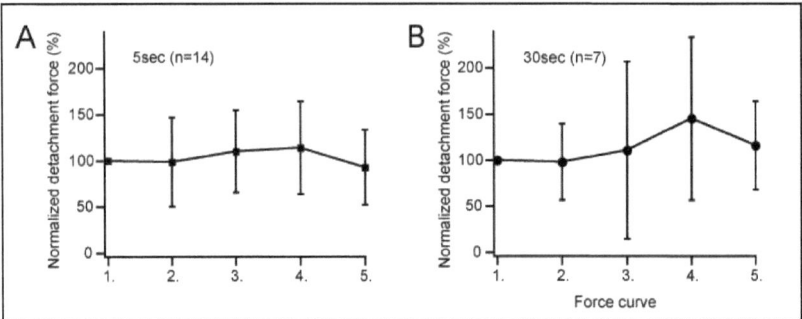

Fig. S1. Force history. Presented are mean detachment forces ± SD detected for several cells. Number of analysed cells are annotated. In independent experiments two different contact times were applied (5 sec (A) and 30 sec (B)).

A.2.2 Testing the mechanical stability of Col

To test wether Col was disrupted by high cell detachment forces, Col was imaged by AFM after recording F-D cycles. Col matrix were apparently not to affected by the detachment process.

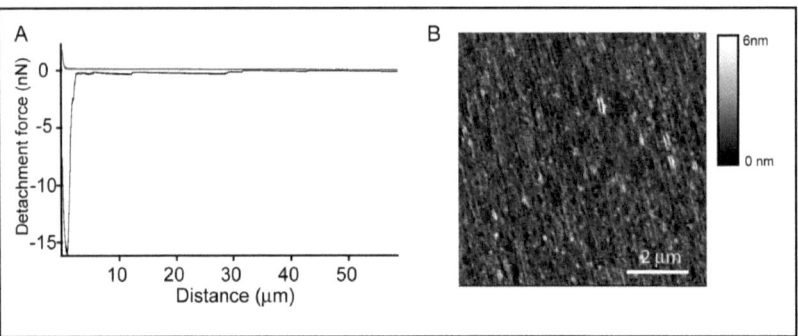

Fig. S2. Testing the mechanical stability of collagen type I matrices. (A) Detachment force curve for a strongly adherent cell (detachment force > 15 nN). (B) AFM topograph showing Col after several F-D cycles.

A.2.3 Statistics of SCFS in chapter 4.

Time (sec)	CHO-A2 n	CHO-A2 # cells	CHO-WT n	CHO-WT # cells	CHO-A2 + 10 µM Y27632 n	CHO-A2 + 10 µM Y27632 # cells	CHO-A2 + 20 mM BDM n	CHO-A2 + 20 mM BDM # cells
5	209	50	65	11	41	14	45	18
10	39	12	-	-	-	-	-	-
30	78	20	30	11	-	-	-	-
60	31	15	-	-	-	-	-	-
120	43	23	31	11	31	11	14	10
180	19	14	-	-	-	-	-	-
240	14	13	-	-	-	-	-	-
300	33	19	16	9	13	10	24	12
600	16	14	-	-	-	-	-	-

n: number of force curves

Table S2. *Number of cells (# cells) analysed and number of force curves (n) generated.*

A3 Supplementary information for chapter 5

A3.1 Characterisation of Col/pdCol coated thermanox discs

In addition to coated mica discs, MC3T3-E1 cells were cultured on Col and pdCol coated *thermanox* discs.

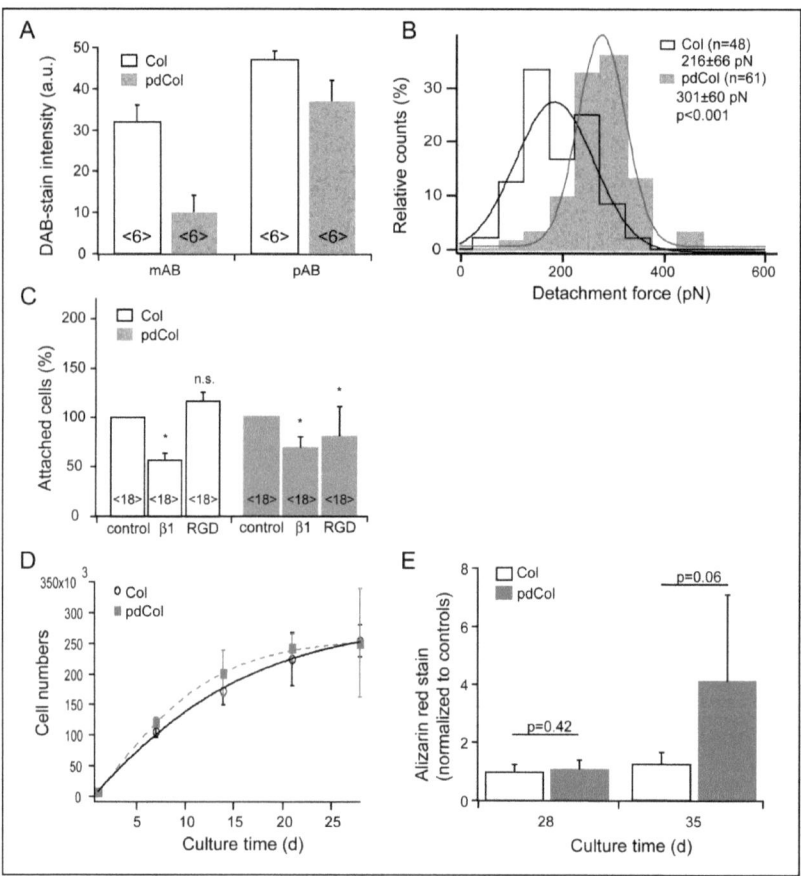

Fig. S3. Characterisation of Col and pdCol- coated thermanox discs. (A) Binding of mAB and pAB collagen antibodies to Col/pdCol coated thermanox discs. Mean DAB (3,3´Diaminobenzidine) stain intensities ± SD are shown. Numbers in brackets indicate the number of analysed thermanox surfaces. (B) Detachment forces of MC3T3-E1 cells after 5 sec contact with Col/pdCol thermanox measured by SCFS. (C) Attachment of MC3T3-E1 cells to Col/pdCol (thermanox) analysed by washing assays in presence and absence of β_1-integrin blocking antibody and RGD peptide. (D) Cell proliferation and matrix mineralization (E) on Col/pdCol coated thermanox discs. Methods are explained in B10 and B11.

Antibody staining of coated thermanox discs confirmed that structural changes occurred in Col after thermal treatment (1 h, 50 °C) similar as observed on coated mica (Fig. S3 A). Moreover, SCFS revealed higher MC3T3-E1 cell adhesion to pdCol than to Col coated *thermanox* discs (Fig. S3 B), such as observed for collagen coated mica. Washing assays confirmed that cell attachment to pdCol coated *thermanox* was RGD-dependent (Fig. S3 C). In line with experiments performed on Col and pdCol coated mica discs, matrix mineralization was more pronounced after 35 d on pdCol compared to Col (Fig. S3 E), whereas cell proliferation was similar (Fig. S3 D).

B   Experimental procedures

B1   AFM- SCFS

B1.1   Experiments

AFM & equipment

For SCFS a *Nanowizard I* (*JPK instruments*) was mounted on top of an inverted light microscope (*Axiovert 200 (Carl Zeiss*, Jena)) equipped with a CCD Camera (*Coolsnap cf*, Diagnostic Instruments Inc.). This setup allowed combined use of AFM and light microscopy (Fig. S4). A special feature of the SCFS setup used was the extended piezo range of 100 μm (*CellHesion* module, *JPK instruments*), which was essential to cell-cell and cell-ECM experiments performed at longer contact times. All experiments were conducted at 37 °C using a temperature-controlled chamber that enabled precise temperature-control in the chamber (±0.1 K) (*BioCell*, JPK instruments)[*].

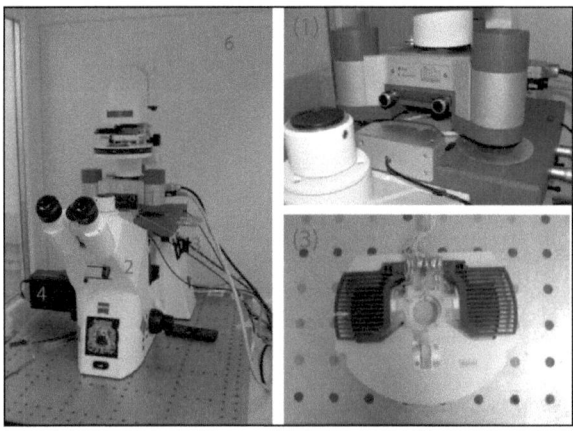

Fig. S4. Photography of the AFM setup. AFM head (1) mounted on an inverted light microscope (2) equipped with a CCD Camera. The sample was mounted in the biocell (3). The setup was placed on a damping table (5) to reduce vibrations transmitted by the building. The whole setup was in an acoustic noise chamber (6).

---

[*] Puech, P.H., Poole, K., Knebel, D. & Muller, D.J. *A new technical approach to quantify cell adhesion forces by AFM. Ultramicroscopy 106, 637-644 (2006).*

The bottom of the *Biocell* consisted of a standard glass coverslip (Ø=24 mm) that was coated with the substrate of interest (ECM proteins or monolayer of cells grown on the coverslip, see table S2). V-shaped, 200 μm long tipless silicium nitride cantilevers with a nominal spring constant of 0.06 N/m (*NP-0, Veeco Instruments*, Woodbury,NY; USA) were used.

Cantilever functionalization for cell attachment

For cantilever functionalization a modified version of the published protocol of *V. Moy's* group was used[*]. NP-O cantilevers (Veeco probes) were cleaned in a plasma cleaner (*Plasma Cleaner/Sterilizer PDC-32G* (Harrick Plasma)) at maximal power for 2 min prior to coating. Then cantilevers were incubated overnight in 50 μl droplets of 0.5 mg/ml BSA-Biotin (see D.6) in $NaCHO_3$ buffer (100 mM, pH 8.6) in a humified chamber at 37 °C. The next day, cantilevers were washed three times in PBS and incubated in 50 μl droplets of 0.5 mg/ml streptavidin in PBS at RT in a humified chamber for 45 min. Thereafter cantilevers were washed three times in PBS and incubated for 45 min in 50 μl droplets of 0.5 mg/ml Concanavalin A (ConA)- Biotin in PBS at RT in a humified chamber. Finally, they were rinsed and stored in PBS. Functionalized cantilevers could be stored at 4 °C for at least two weeks. After use, cantilevers were recycled: for cleaning cantilevers were immersed in 1 M sulfuric acid for 1 h, then rinsed three times with H_2O and plasma-cleaned before re-functionalisation.

Preparation of a cell suspension for SCFS

Cells (table S2 and C) were grown to confluency in cell culture medium. They were harvested using *Accutase*[†] (PAA) or 0.05 % trypsin-EDTA solution 10-15 min before starting experiments. If trypsin was used, it was inactivated by adding trypsin inhibitor. Cells were suspended in CO_2-independent medium (see table S2 and C), centrifuged at 1000 rpm for 3 min and transferred into CO_2-independent medium, either commercially available CO_2-independent medium or DMEM supplemented with 20 mM Hepes. Cells were then kept at 37 °C until they were added into the *Biocell* to start experiments (usually within 10-20 min). For integrin-function blocking

---

[*] Wojcikiewicz, E.P., Zhang, X. & Moy, V.T. *Force and Compliance Measurements on Living Cells Using Atomic Force Microscopy (AFM). Biol Proced Online 6, 1-9 (2004)*.

[†] *Accutase might be more suitable for most experiments than trypsin-EDTA since the effect of adhesion receptor degradation is decreased..*

experiments, cells were pre-incubated with 10 µg/ml blocking antibodies or 100 µM RGD peptide for 30-60 min before performing experiments in presence of inhibitors.

Setting up the experiment

A coverslip with the adhesive substrate (table S2) was mounted into the AFM *BioCell* sample holder (*JPK Instruments*). The surface was rinsed several times with CO_2-independent medium. A Con A-functionalized cantilever was fastened onto the cantilever holder and the holder was mounted into the AFM head. Then the AFM head was set onto the sample stage.

Cantilever calibration

Cantilever spring constants were determined *in situ* prior to every experiment using procedures implemented into the *SPM* software (*JPK Instruments*). The conversion of cantilever deflection into voltage change was determined by pushing the cantilever onto a stiff surface. Then the cantilever spring constant was measured using the so-called thermal noise method[*].

Cell capture

After cantilever calibration ~10^3 suspended cells were injected into the *BioCell* to obtain a sparse distribution on the substrate. Cells have to be captured before they establish firm contact to the substrate. Thus, the following steps should occur as fast as possible, approximately within the first five minutes after cell injection. Immediately after the cells settled down on the substrate, the AFM cantilever was positioned above a single cell by adjusting the position of the AFM head. X-Y positions of AFM head and sample were modified by manually moving the sample stage or the AFM head. The Z position of the AFM head was changed using implemented stepper motors. Cell and cantilever were observed by light microscope throughout the experiment. For cell capture the AFM cantilever was lowered onto a selected cell using the Z piezo. By controlling the relative position of cell and cantilever during approach, it was ensured that the cell was positioned precisely at the extreme end of the cantilever. Usually a contact force of 300-600 pN and a contact time of 2-3 sec were used for cell capture. Thereafter, the cantilever with the bound cell was retracted and brought to a distance of 30-80 µm from the substrate. Before the experiment was started, the cell was allowed to establish firm contact with the cantilever for ~10 min.

---

[*] Hutter, J.L. & Bechhoefer, J. Calibration of atomic-force microscope tips. *Rev Sci Instrum* **64**, 1868-1873 (1993).

Recording force-distance (F-D) curves

The cantilever with the attached cell was brought into contact with the surface at a preset speed until a defined contact force was reached, then the piezo position was maintained constant (=constant height mode). After a defined attachment period (=contact time) the cantilever/cell couple was retracted and the cell was detached from the substrate. During the described process (=F-D cycle), the force (F) was recorded for each piezo position. Between F-D cycles the cell was allowed to rest. The recovery time depended on the contact time used. For short contact times (1-5 sec) recoveries of 1-5 sec were taken, and numerous F-D cycles were performed (10-50). When cells were in contact with the substrate for longer periods, fewer F-D cycles were made, for contact times > 60 sec only one F-D cycle was performed. To avoid structural defects due to repeated contacts, the position on the substrate where contact with the cell occurred was changed between a small set of F-D cycles (1-3). This also compensated for substrate inhomogenities. In table S2 the parameters used for different projects are listed. They were empirically optimized for each experimental setup.

Note: Nonlinearity of piezo movements leads to incorrect piezo height data. This problem can be eliminated by the use of a sensor (e.g. implemented in the JPK Nanowizard) that measures the actual z-piezo position. Thereby non-linearity and hysteresis effects in F-D curves were eliminated..

Furthermore, residual piezo polarization may result in piezo position overshoot after the setpoint is reached ("creep"). This piezo creep may result in a considerable increase of the force applied to the cell, especially at prolonged contact time (>1sec). A "closed-loop" mode was implemented in the JPK instrument. It activates a feedback system that linearizes the piezo movement and minimizes piezo creep. A disadvantage of the feedback is that the feedback mechanism additions noise of the sensor to the data. For single-molecule experiments contact periods are usually in the millisecond range and the contribution of piezo creep is neglectable. To reduce signal noise, for single-molecule experiments the instrument was used in "open-loop".

Project	Cell type on cantilever	Substrate	Medium	Speed (µm/sec)	Initial contact force (pN)	Cell detachment prior experiments by	Feedback Mode
Integrin $\alpha_2\beta_1$-collagen (Chapter 1)	CHO-WT CHO-A2	Collagen Type I	DMEM-Hepes	2.5	1.0	Trypsin-EDTA	Closed-loop
Integrin $\alpha_2\beta_1$-collagen (SMM) (Chapter 1)	CHO-WT CHO-A2	Collagen Type I	DMEM-Hepes	2.5-25	0.1-0.2	Trypsin-EDTA	Open-loop
Adhesion to Col/pdCol (Chapter 2)	MC3T3-E1	Col/pdCol native and heat denatured collagen type I	CO_2-independent medium (Gibco)	5	1.5	Accutase	Closed-loop
Effect of BCR/ABL on cell adhesion (Chapter 3)	32D-V and 32D-BCR/ABL	BMSC (M2-10B4) (80% confluent cell layer on glass) Collagen type I FN	RPMI-Hepes	10	1.5	none (suspension cells)	Closed-loop

Table S3. Overview about experimental setups/parameters used.

B1.2   Data analysis –Overall cell adhesion

Raw data files containing force and measured piezo height data (.txt) were imported into an *Igor Pro 5* procedure window using automated procedures written by P.H. Puech and J. Helenius in Igor Pro 5 (Wavemetrics). F-D curves were plotted and corrected as described below (Fig. S5 A, B). In the corrected F-D curves the detachment force corresponding to the minimum force was determined. In a separated procedure j and t events in F-D curves were automatically detected and quantified (Fig. S5 C, D).

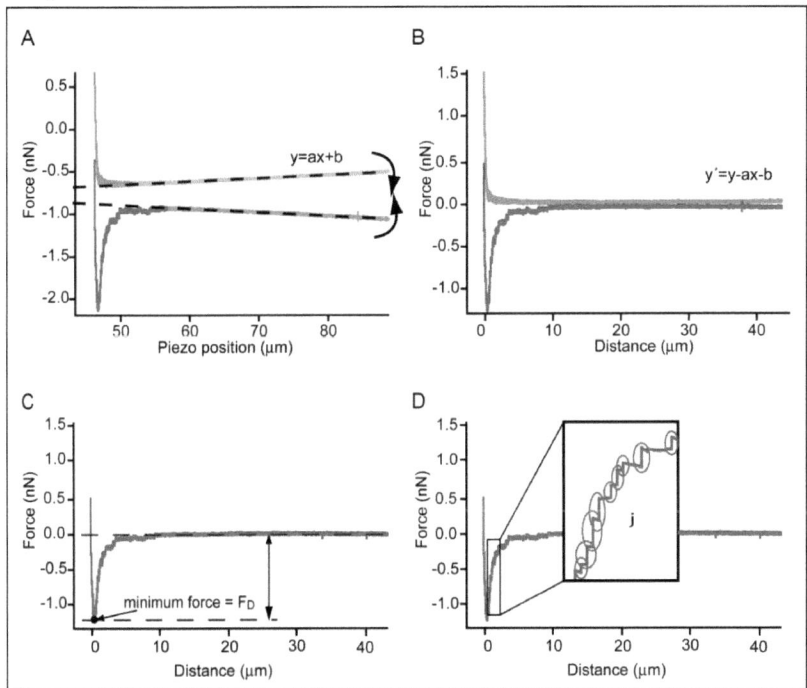

Fig. S5. Analysis of overall cell adhesion and j events in F-D curves after their correction. (A, B) Correction of F-D curves for thermal drift. A line is fitted to the baselines and substracted from the force data. (C) The detachment force is automatically measured by determing the minimum force in the corrected F-D curve. (D) Analysis of j events is done by an automated Igor procedure written by J. Helenius.

Corrections of F-D curves

Unprocessed data F-D curves occasionally have tilted baselines and offsets between trace and retrace baselines (Fig. S5). Trace and retrace baselines tilt in opposite direction are typical for thermal drift. Thermal drift is due to the sensitivity of AFM cantilevers to temperature gradients that are most evident at the beginning of the experiment. Thermal drift is more pronounced when metal-coated cantilevers are used. NP-O cantilevers that were used in the experiments are coated asymmetrically with thin layers of chrome and gold. As these metal layers have different thermal expansion coefficients, temperature changes cause the cantilever to bend. In F-D curves thermal drift is easily corrected by substracting a line (fitted to the baselines) from the force data (Fig. S5). However, high thermal drift affects the performance of the experiment since the contact force cannot be precisely controlled. In such cases the system had to be equilibrated for a longer period. In addition, an offset between approach and retraction baselines due to the hydrodynamic drag may appear, especially for high pulling speeds. When the cantilever is moved through the medium, hydrodynamic forces act opposed to the cantilever movement. Hydrodynamic drag forces increase with the viscosity of the medium, the speed of the cantilever (Fig. S6) and cantilever proximity to the surface. Far from the surface, the drag force is equal to half of the force separating trace and retrace baselines. At low pulling speeds hydrodynamic drag is neglectable, whereas for high pulling speeds e.g. for NP-O smaller 10 µm/sec the hydrodynamic drag force reaches the magnitude of single-molecule binding strength[*] (~ 50 pN, Fig. S6). Thus, corrections for the hydrodrag were only applied for single molecules experiments since high pulling speeds were used (see B1.3).

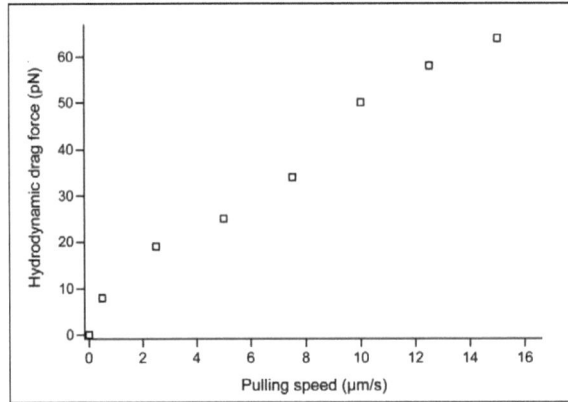

Fig. S6. Hydro-dynamic drag forces versus pulling speed. The hydrodynamic drag force was measured as half of the offset between approach and retraction baselines in F-D curves recorded at different pulling speeds. A NPO cantilever with a bound cell was used.

---

[*] Alcaraz, J. et al. Microrheology of human lung epithelial cells measured by atomic force microscopy. *Biophys J 84, 2071-2079 (2003).*

Janovjak, H., Struckmeier, J. & Muller, D.J. Hydrodynamic effects in fast AFM single-molecule force measurements. *Eur Biophys J 34, 91-96 (2005).*

Other artifacts

In some cases loose objects (e.g. other cell, or dirt particle) floating in proximity or above of the cantilever disturb the cantilever movement or the laser path. In these cases non-linear baselines are observed (Fig. S7, inset). These hinder the precise measurement of F_D and are therefore discarded.

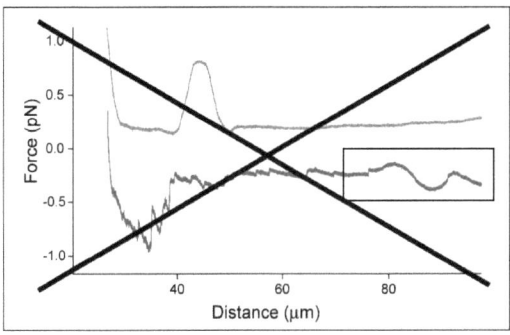

Fig. S7. Typical discarded F-D curve. Apparently an object in the medium interfered with cantilever movement during approach and retraction. Since the baseline is difficult to determine, such F-D curves are discarded.

B1.3 Modified setup for single-molecule measurements

Data acquisition

To perform single molecule measurements, interactions between cell and adhesive substrate have to be reduced such that in the majority of F-D curves only a single receptor/ligand bond unbinds upon cantilever retraction (Fig. S8). This was achieved by choosing a short time period (100-500 msec) and by applying only a small contact force (typically 100-200 pN).

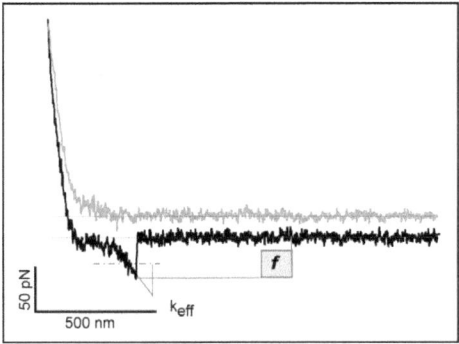

Fig. S8. Representative F-D curve recorded in single-molecules experiments. The F-D curve was recorded with a CHO-A2 cell on collagen type I. Contact time 200msec, contact force 250pN.

The probability that a single unbinding event occurs in repeated F-D curves is -based on Poisson statistics- given by[*]:

$$P(N_b = 1/N_b > 0) = \frac{\lambda}{\exp(\lambda) - 1} \qquad Eq.\ S1$$

where N_b is the number of bonds and λ the frequency of binding events.

Thus, given the case that in only 30 % of the F-D curves a force peak appears, the probability of detecting a single-unbinding event is 86 %.

Since interactions occur under the described conditions rarely, >300 F-D curves can be collected per cell without causing any apparent (as observed by light microscopy) damage to the cell.

F-D cycles were performed at pulling speeds ranging between 0.5–25 µm/sec. By varying the cantilever retraction speed v different $r_{\it eff}$ are applied. $r_{\it eff}$ is defined as follows:

$$r_{\it eff} = v \cdot k_{\it eff} \qquad Eq.\ S3$$

with the effective spring constant $k_{\it eff}$ of the cantilever-cell-substrate-bond system[†]. $k_{\it eff}$ was determined by fitting a line through the final third of the force increase prior to bond rupture (Fig. S8).

Analysis of F-D curves for single-molecule analysis

To obtain statistically significant data, at least 50 F-D curves displaying binding events were collected per pulling speed. To acquire the data, several cells (n > 8) were tested for all pulling speeds. All F-D curves were examined to see if a single rupture event was present. By using an automated software procedure the magnitudes of single rupture events were measured. The F-D data files (.txt) were imported sequentially into an Igor Pro Procedure window, corrected for tilted baselines, baselines were also set to zero. Then the force at the moment of bond rupture was measured (Fig. S 8)

---

[*] Tees, D.F., Waugh, R.E. & Hammer, D.A. *A microcantilever device to assess the effect of force on the lifetime of selectin-carbohydrate bonds.* Biophys J 80, 668-682 (2001).

[†] Zhang, X., Wojcikiewicz, E. & Moy, V.T. *Force spectroscopy of the leukocyte function-associated antigen-1/intercellular adhesion molecule-1 interaction.* Biophys J 83, 2270-2279 (2002).

Correction for the hydrodynamic drag force

Each single force value was corrected to account for the acting hydrodynamic drag force that resulted in a non-neglectable underestimation of unbinding forces at the high pulling speeds that were applied (up to 25 µm/sec). The drag force acting on the cantilever was calculated as follows[*]:

$$F_d = v_{tip} * (6*\pi * \eta * a_{eff})/(h+h_{eff}) \qquad Eq.\ S2$$

with the coefficients a_{eff} and h_{eff} being the effective cantilever area and height, h the cantilever-substrate distance, η the viscosity of the medium. v_{tip} is the tip velocity immediately before bond rupture. The tip velocity is equal to "(piezo position)-(cantilever deflection)". v_{tip} was measured for each unbinding event. a_{eff} and h_{eff} depend on the cantilever geometry and were determined by moving the cantilever through the medium and measuring the hydrodynamic drag for the distance separating cantilever and surface (h). For NP-O cantilevers values of $a_{eff}=57.9$ µm and $h_{eff}=6.3$ µm were obtained. These values were similar to previously reported values for similar cantilevers[†].

Calculation of the bond dissociation rate k_{off} and the barrier width x_u

To generate the dynamic force spectrum, mean rupture forces f_m (here interpreted as binding strength) were calculated for each pulling speed. f_m were plotted versus the logarithm of the corresponding loading rate $\ln(r_{eff})$ (see 3.3)[‡]. Since recorded data (chapter four) were mainly normally distributed, there was no significant difference of using mean, most probable or median rupture force. Rupture force data were fitted using *Eq. S4* and the bond dissociation rate k_{off} and the barrier width x_u (see chapter 3) were extracted (Igor Pro software).

$$f_m = \frac{k_B T}{x_u} \cdot \ln(\frac{x_u}{k_B T \cdot k_{off}}) + \frac{k_B T}{x_u} \cdot \ln(r_{eff}) \qquad Eq.\ S4$$

with the Boltzman constant k_B (1.3806504*10^{-23} J/K) and the *T* the absolute temperature (310 K).

---

[*] Alcaraz, J. et al. *Correction of Microrhelogical Measurements of Soft Samples with Atomic Force Microscopy for the Hydrodynamic Drag on the Cantilever. Langmuir 18, 716-721 (2002).*

[†] Janovjak, H., Struckmeier, J. & Muller, D.J. *Hydrodynamic effects in fast AFM single-molecule force measurements. Eur Biophys J 34, 91-96 (2005).*

[‡] Evans, E. & Ritchie, K. *Dynamic strength of molecular adhesion bonds. Biophys J 72, 1541-1555 (1997).*

Analysis of single membrane tethers

The force of single tethers t were measured by Yi-Ping using a procedure written in *Igor Pro* by Jonne Helenius. t was plotted versus the pulling speed and fitted with following equation:

$$f = f_0 + 2\pi\eta_{eff} v \qquad Eq.\ S5$$

where f is the tether force at finite velocity, v the tether formation velocity (=pulling speed), f_0 the tether force at v=0 and η_{eff} the effective viscosity.

A linear relationship between force and tether extractions speed (i.e. pulling speed) is expected due to both membrane viscosity and friction between membrane and cytoskeleton[*].

---

[*] Shao, J.Y. & Hochmuth, R.M. *Micropipette suction for measuring piconewton forces of adhesion and tether formation from neutrophil membranes.* Biophys J 71, 2892-2901 (1996).

Hochmuth, R.M. & Marcus, W.D. *Membrane tethers formed from blood cells with available area and determination of their adhesion energy.* Biophys J 82, 2964-2969 (2002).

B2 Preparation of adhesive ECM substrates

Collagen type I matrices (Col)

Whereas collagen fibrillogenesis *in vivo* is a complex process involving cells and numerous proteins, collagen fibrillogenesis *in vitro* occurs as a straightforward entropy-driven process[*]. This enables the *in vitro* preparation of pure fibrillar collagen coatings. Concentrated collagen type I solutions can be obtained by acid extraction from collagen-rich tissues, such as skin or tendon. By neutralizing the acidic collagen solution in physiologic buffer solutions, *in vitro* fibril growth can be initiated. The group of Prof. D. Müller has developed several years ago a protocol by which ultra-thin, two-dimensional and highly ordered collagen type I matrices (Col) can be produced on mica surfaces[†]:

To allow mounting into the AFM *Biocell* chamber, mica discs (Ø=4 mm) were glued onto glass cover slips using an optical adhesive (Optical adhesive OP-29, Dymax corporation, Torrington, USA). Then 30 µl of buffer (200 mM KCl, 50 mM Glycine, pH 9.2) were applied onto the freshly cleaved mica discs. Then 1 µl collagen type I solution (2.7 mg/ml) (see D4) was injected into the buffer droplet to yield a final collagen concentration of 90 µg/ml. Samples were incubated overnight at room temperature (RT) in a humid chamber. Prior experiments, surfaces were rinsed with PBS to remove not-attached collagen. The matrices were always kept hydrated.

Partially denatured collagen type I matrices (PdCol)

PdCol matrices were prepared by heating Col matrices for 1 h at 50 °C in a humid chamber. Thereafter matrices were kept at RT for at least 2 h prior to experiments. Col and pdCol surfaces were always kept in buffer solution.

---

[*] *Kadler, K.E., Hill, A. & Canty-Laird, E.G. Collagen fibrillogenesis: fibronectin, integrins, and minor collagens as organizers and nucleators. Curr Opin Cell Biol 20, 495-501 (2008).*

[†] *Jiang, F., Khairy, K., Poole, K., Howard, J. & Müller, D.J. Creating nanoscopic collagen matrices using atomic force microscopy. Microsc. Res. Tech. 64, 435-440 (2004).*

Col/pdCol coated thermanox discs

150 µl (450 µl) of buffer (200 mM KCl, 50 mM Glycine, pH 9.2) were applied onto sterile *thermanox* discs (Ø=13 or 24 mm, depending on their use). Then 5µl (15µl) collagen type I solution (2.7 mg/ml, D4) were injected into the buffer droplet to yield a final collagen concentration of 90 µg/ml. Samples were incubated overnight at room temperature (RT) in a humid chamber. Prior experiments, surfaces were rinsed with PBS to remove not-attached collagen. The matrices were always kept hydrated.

FN coated surfaces

FN was adsorbed to acid-washed coverslips. These were prepared by a 5 h acid wash with 1 M HCl at 50 °C. Subsequently cover slips were rinsed three times with pure ethanol and water and dried. Then coverslips were incubated with 50 µg/ml human plasma FN in PBS containing Ca^{2+} and Mg^{2+} for 90 min at RT. Subsequently, surfaces were rinsed with PBS to remove unbound protein.

B3. AFM Imaging

AFM imaging of protein samples was conducted at RT in buffer solution (PBS) using a *Nanowizard I* (*JPK Instruments*, Berlin, Germany). The piezoelectric scanner had a maximal X-Y scanning range of 100 µm. For AFM imaging in tapping mode (TM) SiO_2 cantilevers (*NPS, Veeco Probes*, Plainview, USA) having a nominal spring constant of 0.06 N/m were used. Drive frequencies close to the resonance frequency of the cantilevers (10-15 kHz) were chosen.

B4. Analyzing mechanical properties of collagen type I matrices

To compare the mechanical stability of native and partially denatured collagen type I matrices (Col/pdCol) (see B2), perpendicularly acting scratching forces were applied to them*. First, an area of 10 x 10 µm² within Col/pdCol matrices was imaged in contact mode AFM at non-destructive contact forces (≈ 50 pN). MSCT cantilevers (k = 0.01 N/m, *Veeco Probes, USA*) were used. Scanning of samples was always performed perpendicular to the collagen fibril direction. Next, 2 x 2 µm² sections within the previously imaged area were scanned by applying increasing contact forces to the AFM stylus; contact forces ranged from 2 nN to 5 nN. Subsequently, the manipulated 10 x 10 µm² area was re-imaged at minimal force (≈ 50 pN) to evaluate the structural changes. For quantification the force at which the AFM stylus started to induce structural deformations (=F_{damage}) was determined. Eight different Col and pdCol matrices, prepared in independent sample preparations, were analysed in total.

B5. Immunostaining of Col/pdCol matrices

Col/pdCol assembled on mica

Col and pdCol matrices were assembled on Ø=10 mm mica discs. Matrices were fixed with 4 % paraformaldehyde (PFA) in PBS for 15 min. After washing with PBS, Col/pdCol surfaces (except negative controls) were incubated for 45 min at RT with 5 µg/ml primary antibodies raised against collagen type I. As primary antibodies a murine monoclonal antibody (mAB) that recognized not-denatured collagen type I and additionally a rabbit polyclonal antibody (pAB) were used (see D6). Antibodies were diluted in 0.5 % bovine serum albumin (BSA) in PBS. After washing three times with PBS, all samples were incubated for 45 min with 3 µg/ml fluorescently labeled secondary antibodies in 0.5 % BSA/PBS. FITC-conjugated anti-mouse IgG and TRITC-conjugated anti-rabbit IgG were used (D6). After washing twice in PBS and a final washing step in ddH$_2$O for salt removal, FITC labeled reference beads (*Calibrite^{TM}3, BD Biosciences*) were applied onto the surfaces. Samples were mounted upside down onto glass cover slips using an anti-bleaching reagent (*ProLong Gold, Invitrogen*). Col/pdCol matrices were examined by confocal microscopy (*LSM 510 Meta, Zeiss, Göttingen, Germany*) using identical imaging parameters. To quantitatively compare mAB binding to Col/pdCol, surfaces were incubated solely with mAB and

---

* Friedrichs, J., Taubenberger, A., Franz, C.M. & Muller, D.J. *Cellular remodelling of individual collagen fibrils visualized by time-lapse AFM. J Mol Biol* 372, 594-607 (2007).

FITC-conjugated anti-mouse. Then mean fluorescence intensities of respective confocal images were determined using *Image J* software (*National Institutes of Health*).

Col and pdcol assembled on *thermanox* discs could not be analysed using fluorescently labelled secondary antibodies, because *thermanox* discs exhibited a strong autofluorescence. Instead 3 µg/ml HRP-conjugated secondary antibodies in 0.5 % BSA/PBS were used. After washing them, matrices were then incubated for 15 min with 3,3´Diaminobenzidine (DAB). *Thermanox* discs were scanned on a standard office scanner and staining intensities were quantified at three different spots per disc using *Image J* software. In total six different surfaces (Col/pdCol) were analysed in two independent experiments.

B6   Immunostaining of cells

Cells were seeded in cell culture medium onto the respective substrate (~25 000 cells /cm^2) for a certain time period. Then the cell layer was washed with PBS (w/ Mg^{2+}, Ca^{2+}), fixed in 4 % PFA/PBS (20 min) and permeabilized with 0.2 % Triton X-100 in PBS for 5min. Next, samples were incubated for 45 min with 5 µg/ml primary in 0.5 % BSA/PBS. After washing three times with PBS, samples were stained with 5 µg/ml fluorescently labelled secondary antibodies, 1µg/ml fluorescently-conjugated phalloidin and 1 µg/ml DAPI for 45 min in 0.5 % BSA/PBS. After washing twice in PBS and a final washing step in ddH$_2$O for salt removal, samples were mounted upside down on glass cover slips using an anti-bleaching reagent. Confocal microscopy images were acquired using a *LSM 510 Meta (Zeiss)*.

B7   Analysis of cell migration and spreading

Col/pdCol matrices were prepared on mica discs (Ø=10 mm) that had been glued onto the glass bottom of a 34 mm Petri dish. Surfaces were carefully rinsed with cell culture medium before 2 ml cell culture medium containing ~5x10^4 MC3T3-E1 cells were added into the *Petri* dish. The *Petri* dish was mounted into a live cell chamber on top of an inverted light microscope (*Axiovert 200M, Zeiss*). The system had been equilibrated to 37 °C and 5 % CO$_2$. Phase contrast images at 10x magnification were recorded every 30 sec over a total period of 90 min. Image sequences were viewed in *Image J* and the percentage of spread cells was determined within the first hour after seeding. Furthermore single cells were tracked using *Image J* (*manual tracking plugin*). From the trajectories of single cells their mean migration speeds were calculated. At least five different movies for each, Col and pdCol matrices, were analysed.

B8   FAK phosphorylation at tyr 397 in MC3T3-E1 cells on Col/pdCol

Col/pdCol matrices were prepared in 6-well dishes as described, washed twice with PBS and once with alpha-MEM. Then ~2.5 x 10^5 MC3T3-E1 cells were seeded onto the matrices. Cells were incubated for 45 and 90 min at 37 °C and 5 % CO_2. Then cells were lysed in 100 µl SDS sample buffer (62.5 mM Tris-HCl (pH 6.8), 2 % (w/v) SDS, 10 % glycerol, 50 mM DTT, 0.01 % (w/v) bromophenol blue) and incubated for 5 min at 95 °C. Cell lysates were analysed by western blot analysis. An antibody recognizing phosphorylated tyr397 was used for detection (see *D6*).

B9   Washing assays

MC3T3-E1 cells were harvested with *Accutase* and transferred into CO_2-independent, serum-free medium. Cells were pre-incubated- when needed- with 10 µg/ml blocking antibodies (D6) or RGD peptide (100 µM) for 30 min prior to experiments. Then ~75000 MC3T3-E1 cells in 150 µl were seeded onto Col/pdCol coated *thermanox* discs placed in 48well plates. After a 30 min-attachment period at 37 °C, loosely and weakly attached cells were removed by carefully rinsing wells twice with 450 µl medium using a 1 ml-pipette (*Eppendorf*). Then 450 µl PBS (with 0.5 mM Mg^{2+} and 1.8 mM Ca^{2+}) were added carefully. Wells were set upside down onto a layer of tissue paper sheets to soak off all liquid. Wells were kept at –80 °C until numbers of attached cells were quantified using the *CyQuant* Proliferations Assay Kit from Molecular Probes.

B10   Cell proliferation assay

Cell numbers were quantified after 7, 14, 21 and 28 days after seeding using the CyQuant Cell Proliferation Assay Kit from Molecular Probes (Invitrogen). After medium removal cells were rinsed with PBS and stored at -80 °C until analysis was performed. Then cells were lysed in cell lysis buffer provided by the manufacturer. Cell lysates were appropriately diluted and 50 µl were transferred into a 96-well plate. Then 50 µl of 2x concentrated *CyQuant* DNA binding dye were added. Fluorescence was detected at 530 nm using a plate reader (*SpectraFluorPlus*, Tecan, Crailsheim, Germany). A cell suspension of known cell concentration was used as standard.

B11   Matrix mineralisation- Alizarin red stain

MC3T3-E1 cells were cultured for 28 and 35 and 42 days on Col and pdCol. Osteogenic differentiation was induced after 1 day by replacing normal culture medium by medium supplemented with osteo-inductive compounds (See C). Controls were furtheron grown in normal cell culture medium. After the respective culture period, medium was removed and cells were washed twice with PBS and fixed for 15 min in 10 % formaldehyde at RT. Then samples were washed carefully with dH$_2$O and incubated for 20 min with 2 % *Alizarin red S* solution (pH 4.1 - 4.3) under gentle shaking. Thereafter, samples were rinsed three times with ddH$_2$O water until no further dye was released. Samples were dried and kept frozen at –20 °C until dye extraction. Quantification of calcified matrix was performed following the protocol from Gregory et al.[*]. Briefly, 200 µl of 10 % (v/v) acetic acid were added to each well and incubated for 30 min at RT on a rotary plate. Then the cell layer was scraped off and transferred in acid solution to a 500 µl reaction tube. After vortexing for 30 sec, the reaction tubes were heated at 85 °C for 10 min. Thereafter samples were cooled on ice and centrifuged at 16000g for 15 min. Then 125 µl of the supernatant were taken and neutralized with 50 µl 10 % (v/v) ammonium hydroxide. The pH of the supernatant was measured to ensure that it ranged between 4.1 and 4.3. 150 µl of the samples were transferred into 96-well plates and read at 405 nm using a plate reader.

B12   Flow Cytometry

A solution of PBS (PBS w/ Mg^{2+}, Ca^{2+}) containing 2 % BSA and 0.02 % Sodium-azide was prepared and cooled on ice. Then cells were harvested using *Accutase* and transferred into the prepared BSA solution. After washing once, a cell suspension of 10^6 cells/ml was prepared and kept on ice. 100 µl of cell suspension were incubated for each sample in dublicate. Samples were incubated with primary antibodies (20 µg/ml) (D6) for 1 h. After washing twice with BSA-PBS cells were incubated with secondary antibodies (10 µg/ml) (D6) for 45 min (if non-labelled primary antibodies had been used before). Controls were solely incubated with fluorescently labelled secondary antibodies or fluorescently labelled unspecific antibodies. Samples were washed three times with ice-cold PBS (w/ Mg^{2+}, Ca^{2+}) and fixed in 4 % formalin (in PBS) for 20 min at RT.

---

[*] *Gregory, C.A., Gunn, W.G., Peister, A. & Prockop, D.J. An Alizarin red-based assay of mineralization by adherent cells in culture: comparison with cetylpyridinium chloride extraction. Anal Biochem 329, 77-84 (2004).*

Finally cells were washed with PBS and fluorescence staining of 20000 cells was analysed in a flow cytometer (BD Biosciences).

B13 SDS-PAGE and western blot analysis

30 µl sample were loaded onto 8 % polyacrylamide (PA) gel, proteins were separated for 60 to 90 min at 110 V. Proteins were thereafter transferred onto nitrocellulose membranes (*Schleicher&Schuell*, Dassel, Germany). After blocking for one hour in milk buffer (TBS-Tween 0.1 %, 5 % skim milk) membranes were incubated overnight with primary antibodies. After washing, membranes were incubated with secondary HRP-conjugated antibody (D6). After a final washing step in TBS (Tris-buffered solution)-Tween (0.1 %) ECL (*Biorad*) was added and chemiluminescence was detected on a photosensitive film (both from *Amersham Bioscience*). Loading controls were performed with anti-β-tubulin or vinculin antibodies (D6) after stripping the antibodies off the membrane.

B14. Reverse transcription and quantitative real-time PCR

(by the collaborator Fernando Fierro)

Quantitative real-time PCR was performed using TaqMan® Gene Expression Assays (Applied Biosystems, Foster City, CA, USA); 1 µL cDNA synthesized from total RNA was used. Pre-designed primer sets for mouse itgb1 were purchased from Applied Biosystems. Amplification conditions were: one initial cycle of 50 °C for 2 min plus 95 °C for 10 min followed by 40 cycles of 95 °C for 15 sec and 60 °C for 1 min. Itgb1 expression was normalized by comparison with the expression of the housekeeping gene glyceraldehyde-3-phosphate-dehydrogenase (Applied Biosystems).

B15. Statistical analyses

All data were generated by at least three independent experiments. Non-Gaussian distributed datasets were compared by a non-parametric two-sided significance test (Mann-Whitney test) using *Instat* software. Significance of Gaussian distributed data sets was tested by t-tests. Datasets were considered as significantly different if p-values were smaller than 0.05.

C. Cell culture

CHO cells. Chinese hamster ovary cells. CHO-WT and CHO-A2 (stably expressing $\alpha_2\beta_1$-integrin) cells were a kind gift from J. Heino. Cells were grown for continuous passaging in alpha-MEM supplemented with 10 % FCS, 1 % L-glutamine, penicillin (10000U/ml) and streptomycin sulfate (10000 µg/ml). Medium of CHO-A2 cells was further supplemented with 0.4 µg/ml geneticin. 2-3 times a week, upon reaching confluency, cells were detached from the tissue culture flasks by trypsin-EDTA (0.05 %) and passaged at a ratio of 1:10 into a new culture flask.

Saos-2 cells. Human osteogenic sarcoma cells. Saos-WT and Saos-A2 cells were a kind gift from J. Heino. Cells were grown for continuous passaging in DMEM supplemented with 5 % FCS, 1% L-glutamine, penicillin (10000 U/ml) and streptomycin sulfate (10000µg/ml). Medium of Saos-A2 cells was further supplemented with 0.2 µg/ml geneticin. 2-3 times a week cells were passaged at a ratio of 1:10.

M2-10B4. Mouse bone marrow derived stromal cells. M2-10B4 were purchased from the American type culture collection (ATCC, Wesel, Germany) and cultured with RPMI 1640 supplemented with 10 % fetal calf serum (FCS). 2-3 times a week, cells were passaged at a ratio of 1:10.

Preparation of M2-10B4 monolayers for SCFS. One day before AFM experiments, ~150000 M2-10B4 cells were seeded onto sterile glass coverslips within 6-well dishes to obtain a cell layer of about 75 % confluency for experiments

32D cells. Mouse bone marrow cells (myeloid progenitors). 32D control cells were retrovirally transfected with empty retroviral Mig vector (32D-V) or p210$^{BCR/ABL}$ (32D-BCR/ABL). Cells were kindly provided by A. Neubauer (Marburg, Germany). 32D cells were cultured in suspension using RPMI 1640 + 10 % FCS. 32D-V culture medium was supplemented with 1 U/ml recombinant mouse interleukin 3 (IL-3) (Strathmann Biotec, Hamburg, Germany). 32D-BCR/ABL cells were growth factor independent.

MC3T3-E1. Mouse embryo calvaria cells. MC3T3-E1 cells were a kind gift from B. Hoflack (*Biotec, TU Dresden*). Cells were grown for continuous passaging in alpha-MEM supplemented with 10 % FCS, 1 % L-glutamine, penicillin (10000 U/ml) and streptomycin sulfate (10000 µg/ml). 2-3 times a week cells were passaged (1:3).

4-weeks culture of cells on Col/pdCol matrices. MC3T3-E1 cells were seeded at a density of ~3000 cells/cm^2 onto Col/pdCol coated 13mm *thermanox* discs in cell culture medium. The next day, medium was replaced by cell culture medium supplemented with osteoinduction medium complements (10 mM β-glycerophosphate, 10 µM 1α,25-dihydroxycholecalciferol (Vitamin D3), 50 µM ascorbic acid). Controls were grown in cell culture medium lacking osteoinductive complements. Media were renewed three times a week.

Recipes for cell culture media used in AFM-SCFS

CO_2-independent medium (DMEM-Hepes) (Chapter 4)
For a total volume of 1l:

DMEM	10.1 g
NaHCO$_3$	350 mg
HEPES	4.77 g
Penicillin/streptomycin (100x)	10 ml
dH$_2$O	990 ml

→ adjust pH to 7.2
→ sterile-filter medium and keep at 4°C

CO_2-independent medium w/o Mg^{2+} and Ca^{2+}

(equivilant to media formulation of DMEM-Hepes, but w/o Mg^{2+} and Ca^{2+})

For a total volume of 1L:

Hank´s balanced salt solution	100 ml
D-Glucose	3500 mg
MEM aminoacids solution	20 ml
MEM vitamin solution	40 ml
L-Serine	42 mg
L-Glutamine	584 mg
L-Glycine	30 mg
$NaHCO_3$	350 mg
Penicillin/Streptomycin	10 ml
dH_2O	830 ml

→ adjust to pH 7.2 using NaOH

→ sterile-filter medium and keep at 4°C

Note: All buffers and solutions are prepared using ultrapure deionized H_2O (18.2 $M\Omega$)

D   Materials

D1.   Chemicals and media ingredients

1a,25-dihydroxycholecalciferol (vitamin D3), Sigma

Accutase, PAA

Alizarin Red S, Sigma

Alpha-MEM, Gibco (Invitrogen, Karlsruhe, Germany).

Ascorbic acid, Sigma

β-glycerophosphate, Sigma

CO_2-independent medium, Gibco

DMEM (1x, liquid), Gibco

DMEM (powder, 1000mg/l Glucose, w/Pyruvate, w/o $NaHCO_3$), Gibco

FCS (fetal calf serum) (FBS), Gibco

Hank´s Balanced Salt Solution (10x HBSS, w/o Ca^{2+}, Mg^{2+}), Gibco

HEPES (4-(2-hydroxyethyl)-1-piperazineethanesulfonic acid), Carl Roth GmbH

L-Glutamine, Gibco

L-Glycine, Sigma

L-Serine, Fluka

MEM vitamin solution (100x), Gibco

MEM amino acids solution (50x), Gibco

$NaHCO_3$, Sigma

PBS, phosphate buffered saline containing Ca^{2+}, Mg^{2+}, Biotec media kitchen

PBS, phosphate buffered saline without Ca^{2+}, Mg^{2+}, Biotec media kitchen

penicillin/streptomycin (100x), Gibco

RPMI 1640, Gibco

tryspin-EDTA (0.5%), Gibco

trypsin inhibitor, Sigma

D2. Plastic equipment

6-well dishes, Nalgene Nunc

Steritop filter units (500 ml), Millipore

Sterile plastic pipettes, Greiner

Thermanox discs, Nalgene Nunc, Thermo Fisher Scientific, Waltham, USA

Tissue culture flasks, Nalgene Nunc

D3. AFM cantilevers

NP-O, Vecco probes, Plainview, USA

MLCT, Veeco probes

MSCT, Veeco probes

D4. ECM proteins

Bovine skin collagen type I (Purecol), Inamed Biomaterials (Fremont, USA).

Human plasma FN, Roche Pharma AG, (Basel, Switzerland).

D5. Proteins for cantilever functionalization

Bovine serum albumin, biotinamidocaproyl-labeled, Sigma

Streptavidin, Sigma

Concanavalin A, biotin conjugated, Sigma

D6. Antibodies & blocking peptides

Integrin function-blocking antibodies

anti-β_1 integrin (clone Ha2/5), *BD Biosciences*, New Jersey, *USA*

anti-β_3 integrin (clone 2C9.G2), *Biolegend,* San Diego, USA anti-α_V integrin (Clone RMV-7), Biolegend

anti-$\alpha_5\beta_1$ integrin (Clone BMB5), *Millipore*, Billerica, USA

Blocking peptides

Linear RGD peptide (GRGDSPK), Sigma

λ229ox (CTRKKHDNAQC), Eurogentec Deutschland GmbH, Köln, Germany

Immunofluorescence

mouse monoclonal anti-collagen type I (clone COL-1), Sigma

rabbit polyclonal anti-collagen type I, Abcam, Cambridge, UK

Western Blots

Anti-FAK ptyr397, New England Biolabs GmbH, Frankfurt, Germany

Anti-human α_2-integrin, Millipore

Anti-rat β_1-integrin, BD Biosciences

Anti-hamster β_1-integrin, kind gift from J. Heino, Turku, Finland

Anti-human vinculin, Sigma

FITC- conjugated anti-mouse IgG, Jackson Immunoresearch, West Grove, USA

TRITC/FITC- conjugated anti-rabbit IgG, Jackson Immunoresearch

Horseradish peroxidase (HRP)-conjugated anti-mouse/rabbit IgG, Amersham Biosciences

Acknowledgements

Last not least I would like to acknowledge all those people who made this work possible:

Firstly my supervisor Prof. Dr. Daniel Müller, for taking me in his team, his support and great ideas and suggestions.

Special thanks to Dr. Clemens Franz, Dr. Pierre-Henri Puech and Dr. Jonne Helenius for all their help with my project, great discussions, and for reviewing my thesis.

Thanks to all my labmates from the Mueller lab, for being always helpful and creating a amicable atmosphere in the lab.

My collaboration partners Dr. Fernando Fierro, Dr. Thomas Ilmer, Prof. Dietmar Hutmacher, Dr. Maria Woodruff and Ms. Huifen Bai.

The JPK team for technical support.

Thanks to Thomas, my family and my friends for their support.

Die VDM Verlagsservicegesellschaft sucht für wissenschaftliche Verlage abgeschlossene und herausragende

Dissertationen, Habilitationen, Diplomarbeiten, Master Theses, Magisterarbeiten usw.

für die kostenlose Publikation als Fachbuch.

Sie verfügen über eine Arbeit, die hohen inhaltlichen und formalen Ansprüchen genügt, und haben Interesse an einer honorarvergüteten Publikation?

Dann senden Sie bitte erste Informationen über sich und Ihre Arbeit per Email an *info@vdm-vsg.de*.

Sie erhalten kurzfristig unser Feedback!

VDM Verlagsservicegesellschaft mbH
Dudweiler Landstr. 99
D - 66123 Saarbrücken
Telefon +49 681 3720 174
Fax +49 681 3720 1749
www.vdm-vsg.de

Die VDM Verlagsservicegesellschaft mbH vertritt

Printed by Books on Demand GmbH, Norderstedt / Germany